Lecture Notes in Mathematics   2106

Editors-in-Chief:
J.-M. Morel, Cachan
B. Teissier, Paris

For further volumes:
http://www.springer.com/series/304

Saint-Flour Probability Summer School

The Saint-Flour volumes are reflections of the courses given at the Saint-Flour Probability Summer School. Founded in 1971, this school is organised every year by the Laboratoire de Mathématiques (CNRS and Université Blaise Pascal, Clermont-Ferrand, France). It is intended for PhD students, teachers and researchers who are interested in probability theory, statistics, and in their applications.

The duration of each school is 13 days (it was 17 days up to 2005), and up to 70 participants can attend it. The aim is to provide, in three high-level courses, a comprehensive study of some fields in probability theory or Statistics. The lecturers are chosen by an international scientific board. The participants themselves also have the opportunity to give short lectures about their research work.

Participants are lodged and work in the same building, a former seminary built in the 18th century in the city of Saint-Flour, at an altitude of 900 m. The pleasant surroundings facilitate scientific discussion and exchange.

The Saint-Flour Probability Summer School is supported by:

– Université Blaise Pascal
– Centre National de la Recherche Scientifique (C.N.R.S.)
– Ministère délégué à l'Enseignement supérieur et à la Recherche

For more information, see back pages of the book and
http://math.univ-bpclermont.fr/stflour/

Jean Picard
Summer School Chairman
Laboratoire de Mathématiques
Université Blaise Pascal
63177 Aubière Cedex
France

Krzysztof Burdzy

Brownian Motion
and its Applications
to Mathematical Analysis

École d'Été de Probabilités
de Saint-Flour XLIII – 2013

 Springer

Krzysztof Burdzy
Department of Mathematics
University of Washington
Seattle, WA, USA

ISBN 978-3-319-04393-7   ISBN 978-3-319-04394-4 (eBook)
DOI 10.1007/978-3-319-04394-4
Springer Cham Heidelberg New York Dordrecht London

Lecture Notes in Mathematics ISSN print edition: 0075-8434
                         ISSN electronic edition: 1617-9692

Library of Congress Control Number: 2014931653

Mathematics Subject Classification (2010): 60J65, 60H30, 60G17

Printed on acid-free paper

Springer is part of Springer Science+Business Media (www.springer.com)

*To Janina Burdzy, my mother, who taught me
the foundations of probability*

Preface

Depending on whom you ask, Brownian motion was either discovered or invented by Louis Bachelier in 1900 (according to economists), by Albert Einstein in 1905 (according to physicists), or by Norbert Wiener in 1923 (according to mathematicians). Biologists do not seem to be interested much in taking credit for the discovery; perhaps, they consider it obvious that they get the credit because Brownian motion is named after a biologist, Robert Brown. He observed random motion of pollen grains looking through a microscope in 1827. According to Wikipedia [Wik12a], the history of Brownian motion can be traced to ancient Rome where Lucretius, in 60 BC, wrote a remarkable description of Brownian motion of dust particles and used this as a proof of the existence of atoms. The same article in Wikipedia claims that Brownian motion was "discovered" multiple times and earlier than by people normally given credit for the discovery.

Brownian motion is a good model for a wide range of real random phenomena, from chaotic oscillations of microscopic objects, such as flower pollen in water, to stock market fluctuations. Brownian motion is also a purely abstract mathematical tool which can be used to prove theorems in "deterministic" fields of mathematics. These lecture notes contain an introduction to the applications of Brownian motion to analysis.

I do not think that there is a well-defined area of probability called "applications of Brownian motion to analysis" and I will not try to create such a field in these notes. Instead, I will present a number of diverse applications of Brownian motion to analysis and, more generally, connections between Brownian motion and analysis. All I can hope for is that summer school participants and other readers will be infected by my enthusiasm and will try to find their own research projects in this area.

Probability theory started as a science describing real-life random phenomena and only more recently developed into a branch of mathematics that generates tools which can be used to prove results in other branches of mathematics. Perhaps the best known application of probability to proving deterministic mathematical theorems is known as the "Erdös probabilistic method." According to Wikipedia [Wik12b], "it is a nonconstructive method, primarily used in combinatorics, for proving the existence of a prescribed kind of mathematical object. It works by showing that if one randomly chooses objects from a specified class, the probability that the result is of the prescribed kind is more than zero." Another Wikipedia article [Wik12c] lists a number of applications of probability to proving theorems in the following fields of mathematics: analysis, combinatorics, algebra, topology, geometry, number theory, and quantum theory.

The history of applications of Brownian motion to analysis goes back at least to several papers of Kakutani in the mid-1940s (see [Kak44, Kak45a, Kak45b]). Other early players in the field were Burkholder [Bur76], Davis [Dav75, Dav79a], and Doob [Doo61]. I apologize to all my colleagues who contributed to the field but are not mentioned in my Twitter-like history of the subject. I have unfairly listed only some mathematicians who are independently famous. Since 1980, interactions between Brownian motion and stochastic analysis, on the one hand, and various branches of analysis, on the other hand, have become so diverse and numerous that I do not feel competent to provide an even remotely accurate review.

The topics for these notes were selected either because I did research on them or because I considered them elegant. Chapter 1 contains a very general review of Brownian motion. Chapter 2 is concerned with probabilistic proofs of classical theorems in analysis. All the remaining chapters contain mathematics developed after 1990.

I regret to say that these notes do not present material at the level of rigor that is expected from journal articles or textbooks published in mathematics. It would take several hundred pages to present rigorously all the mathematical tools needed in these notes. I kindly request that the reader consult one or more of several excellent books that provide an introduction to Brownian motion and its relationship to analysis. Personally, I often consult books by Bass [Bas95], Karatzas and Shreve [KS91], Knight [Kni81], Mörters and Peres [MP10], and Revuz and Yor [RY99]. Other books are referred to in the body of the text.

These notes contain previously published material, in the sense of mathematical substance. To large extent, they are a concatenation of many articles, edited to various degree. I am grateful to Mihai Pascu for the permission to reuse material from his articles in these notes.

I would like to express my gratitude to Donald Marshall for very helpful discussions.

I thank Laurent Serlet, the organizer of the Saint-Flour Probability Summer School 2013, and the Scientific Board for the invitation to deliver the lectures. My research was generously supported by the National Science Foundation via Grant DMS-1206276.

Seattle, WA   Krzysztof Burdzy
2013

Notation

The notation will not be perfectly consistent throughout these notes. The following list contains some of the most frequently used symbols.

The sets of all real and complex numbers will be denoted $\mathbb{R}$ and $\mathbb{C}$. The real and imaginary parts of x will be denoted $\operatorname{Re} x$ and $\operatorname{Im} x$. The Euclidean norm in $\mathbb{R}^d$ will be denoted $|\cdot|$ or $\|\cdot\|$.

An open ball with center x and radius r will be denoted $\mathcal{B}(x, r)$. We will identify points $x \in \mathbb{R}^2$ with vectors $\overrightarrow{(0,0), x}$ and complex numbers $x = re^{i\theta}$. The angle between $x = r_x e^{i\theta_x}$ and $y = r_y e^{i\theta_y}$, i.e., $\theta_x - \theta_y$, will be denoted $\angle(x, y)$. We will use the convention that $\angle(x, y) \in (-\pi, \pi]$.

A function f is called Lipschitz with constant c if $|f(x) - f(y)| \leq c|x - y|$ for all x and y. A domain (i.e., an open connected subset of $\mathbb{R}^d$) is called a Lipschitz domain if its boundary can be represented, in a neighborhood of every boundary point, as the graph of a Lipschitz function in some orthonormal coordinate system. The inward unit normal vector at a boundary point $x \in \partial D$ of a domain D will be denoted $\mathbf{n}(x)$ (provided it exists).

For any process Z_t, we will denote the hitting time of a set A by T_A^Z or $T^Z(A)$, i.e., $T_A^Z = \inf\{t \geq 0 : Z_t \in A\}$. The superscript will be dropped if no confusion may arise. The exit time from a set A will be denoted τ_A, that is, $\tau_A = \tau_A^Z = \inf\{t \geq 0 : Z_t \notin A\}$.

Brownian motion will be typically denoted B_t or W_t. Reflected Brownian motion will be often denoted X_t or Y_t.

The distribution of Brownian motion B starting from x will be denoted $\mathbb{P}^x$. For a probability measure μ on $\mathbb{R}$, $\mathbb{P}^\mu$ will denote Brownian motion with the initial distribution equal to μ. The corresponding expectations will be denoted $\mathbb{E}^x$ and $\mathbb{E}^\mu$.

Typically, $\varphi_1, \varphi_2, \ldots$ will denote eigenfunctions of the Laplacian in a Euclidean domain (with Neumann or some other boundary conditions). The eigenvalues will be denoted by $\mu_1 \leq \mu_2 \leq \mu_3 \ldots$, with the convention that $\mu_k \geq 0$.

Contents

Chapter 1
Brownian Motion

For detailed introductions to the mathematical theory of Brownian motion see [KS91, Kni81, MP10, RY99].

We start our tour of the Brownian motion theory with a "paradox." Some physicists define Brownian motion B_t as a process with the following properties. For any function $f : [0, 1] \to \mathbb{R}$ satisfying $f(0) = 0$ and $\sup_{t \in [0,1]} f''(t) < \infty$, we have

$$\mathbb{P}(f(t) - \varepsilon < B_t < f(t) + \varepsilon, t \in [0, 1]) \approx c(\varepsilon) \exp\left(-\frac{1}{2} \int_0^1 (f'(t))^2 dt\right). \quad (1.1)$$

The right hand side of (1.1) is maximized when $f \equiv 0$. It is a common belief that Brownian motion is a reasonably good model for (the logarithm of) market indices such as the Dow Jones Industrial Average.

Paradox: See Fig. 1.1 and ask yourself how this graph can be reconciled with the claim that the constant function is the "most likely" shape of the Brownian trajectory.

Formula (1.1) was proved in [Syt75, Syt77, Syt79] (see also [Nov81] and results on and references to the "Onsager-Machlup function" in [IW89, Sect. VI.9]). The "paradox" is explained by a theorem presented in Sect. 1.6.

Definition 1.1. Brownian motion $\{B_t, t \geq 0\}$ is the unique (in distribution) process with the following properties.

(i) No memory. If $0 \leq t_0 \leq t_1 \leq \dots$ then

$$B_{t_1} - B_{t_0}, B_{t_2} - B_{t_1}, B_{t_3} - B_{t_2}, \dots$$

are independent.
(ii) Invariance. For $s, t \geq 0$, the distribution of $B_{s+t} - B_s$ depends only on t.
(iii) Continuity. With probability 1, $t \to B_t$ is continuous.
(iv) Normalization. With probability 1, $B_0 = 0$. For all $t \geq 0$, $\mathbb{E} B_t = 0$ and $\mathbb{E} B_t^2 = t$.

K. Burdzy, *Brownian Motion and its Applications to Mathematical Analysis*,
Lecture Notes in Mathematics 2106, DOI 10.1007/978-3-319-04394-4_1,
© Springer International Publishing Switzerland 2014

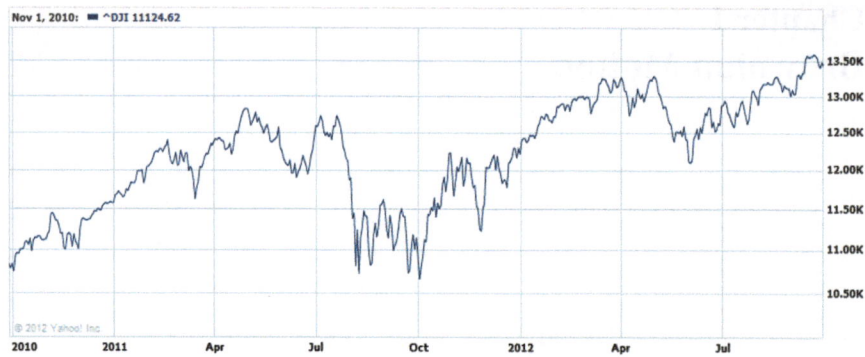

Fig. 1.1 Dow Jones Industrial Average on the logarithmic scale

Proposition 1.2. *With probability 1, t → B_t is nowhere differentiable. See Fig. 1.1.*

1.1   Why Brownian Motion?

Brownian motion belongs to several families of well understood stochastic processes:

1. Markov processes
2. Martingales
3. Gaussian processes
4. Lévy processes

1.1.1   Markov Processes

A simple version of the Markov property says that for $s \geq 0$, the conditional distribution of $\{B_t, t \geq s\}$ given B_s is the same as the conditional distribution of $\{B_t, t \geq s\}$ given $\{B_u, 0 \leq u \leq s\}$.

Markov processes, that is, processes with "no memory," are natural models for many natural phenomena, social phenomena and man-made systems.

On the mathematical side, the theory of Markov processes uses tools from several branches of analysis:

1. Functional analysis (transition semigroups)
2. Potential theory (harmonic functions, Green functions)
3. Spectral theory (eigenfunction expansion)
4. Partial differential equations (heat equation)

Fig. 1.2 Running martingale

Fig. 1.3 Standing martingale

1.1.2 Martingales

See Figs. 1.2 and 1.3. A simple version of the martingale property says that if $0 \leq s \leq t$ then, a.s.,

$$\mathbb{E}(B_t \mid B_u, 0 \leq u \leq s) = B_s.$$

Martingales are processes with "no drift."

There are very few natural phenomena that can be realistically represented by martingales. Martingales are an artificial and very successful mathematical construct.

Semimartingales Y are the only family of processes for which the theory of stochastic integrals of the form $\int_0^t X_s dY_s$ is fully developed, successful and satisfactory.

1.1.3   Gaussian Processes

If $0 \leq t_0 \leq t_1 \leq \dots$ then for any n, the vector

$$(B_{t_0}, B_{t_1}, B_{t_2}, \dots, B_{t_n})$$

has a multidimensional normal (Gaussian) distribution.

1. There are excellent bounds for tails of Gaussian distributions.
2. A Gaussian process is fully characterized by the mean and covariance structure.
3. Gaussian processes can be easily defined on non-ordered (non-time-like) parameter spaces, unlike Markov processes and martingales.
4. Gaussian processes arise naturally in limit theorems and are widely used in modeling of noise.

1.2   Invariance Principle

The invariance principle is a generalization of the Central Limit Theorem. A simple version of the invariance principle is the following. Suppose that $\{R_k, k \geq 1\}$ are i.i.d. with $\mathbb{P}(R_k = 1) = \mathbb{P}(R_k = -1) = 1/2$ and let $S_0 = 0$, $S_k = \sum_{j=1}^{k} R_j$, $k \geq 1$. If B_t is a Brownian motion starting from 0 then

$$\{r^{-1/2} S_{\lfloor tr \rfloor}, t \geq 0\} \to \{B_t, t \geq 0\}$$

in distribution, as $r \to \infty$. See [Bil99] for an excellent introduction to convergence in distribution.

1.3   Multidimensional Brownian Motion

If $B^1, B^2, \dots, B^d$ are independent Brownian motions then $\{(B_t^1, B_t^2, \dots, B_t^d), t \geq 0\}$ is called d-dimensional Brownian motion.

1.4   Itô Integral and Itô Formula

Let $\mathcal{F}_t = \sigma\{B_s, s \leq t\}$ and consider a bounded process X_t such that X_t is $\mathcal{F}_t$-measurable for every $t \geq 0$. Then for $t \geq 0$,

$$\lim_{n \to \infty} \sum_{k=0}^{\lfloor nt \rfloor} X_{k/n} (B_{(k+1)/n} - B_{k/n})$$

exists in L^2 and is denoted $\int_0^t X_s dB_s$.

To state the Itô formula, we generalize the concept of Brownian motion as follows. If $\{B_t, t \geq 0\}$ is a Brownian motion and $\hat{B}_0$ is a random variable independent of $\{B_t, t \geq 0\}$ then we will call $\{\hat{B}_t, t \geq 0\} := \{\hat{B}_0 + B_t, t \geq 0\}$ Brownian motion starting from $\hat{B}_0$.

Theorem 1.3 (Itô Formula). *Suppose that* $f : \mathbb{R} \to \mathbb{R}$ *is twice continuously differentiable and* B_t *is Brownian motion. Then for every* $t \geq 0$, *a.s.,*

$$f(B_t) = f(B_0) + \int_0^t f'(B_s)dB_s + \frac{1}{2} \int_0^t f''(B_s)ds.$$

1.4.1 Brownian Martingales

If X is bounded then $M_t := \int_0^t X_s dB_s$ is an $\mathcal{F}_t$-martingale, that is, $\mathbb{E}(M_t \mid \mathcal{F}_s) = M_s$ for $0 \leq s \leq t$. According to the Martingale Representation Theorem, if M_t is a martingale relative to the filtration $\{\mathcal{F}_t\}_{t \geq 0}$ then there exists an $\mathcal{F}_t$-adapted process X such that $M_t = \int_0^t X_s dB_s$.

1.4.2 Generalizations of the Itô Formula

If $f : \mathbb{R}^d \to \mathbb{R}$ is twice continuously differentiable and B is a d-dimensional Brownian motion then

$$f(B_t) = f(B_0) + \int_0^t \nabla f(B_s)dB_s + \frac{1}{2} \int_0^t \Delta f(B_s)ds, \qquad t \geq 0.$$

If $f : [0, \infty) \times \mathbb{R} \to \mathbb{R}$ is continuously differentiable in the first variable and twice continuously differentiable in the second variable then for $t \geq 0$,

$$f(t, B_t) - f(0, B_0) = \int_0^t \frac{\partial}{\partial x} f(s, B_s)dB_s + \int_0^t \frac{\partial}{\partial s} f(s, B_s)ds$$
$$+ \frac{1}{2} \int_0^t \frac{\partial^2}{\partial x^2} f(s, B_s)ds.$$

1.4.3 Disappearing Terms in the Itô Formula

Suppose that $f : \mathbb{R}^d \to \mathbb{R}$ is twice continuously differentiable, $\Delta f \equiv 0$, and B is a d-dimensional Brownian motion. Then

$$f(B_t) = f(B_0) + \int_0^t \nabla f(B_s)dB_s, \qquad t \geq 0.$$

Suppose that $f : [0, \infty) \times \mathbb{R} \to \mathbb{R}$ is continuously differentiable in the first variable, twice continuously differentiable in the second variable and $\frac{\partial}{\partial x} f(t, x)$ is uniformly bounded. Then for $t \geq 0$,

$$\mathbb{E}\, f(t, B_t) - \mathbb{E}\, f(0, B_0) = \mathbb{E} \int_0^t \frac{\partial}{\partial s} f(s, B_s) ds + \frac{1}{2} \mathbb{E} \int_0^t \frac{\partial^2}{\partial x^2} f(s, B_s) ds.$$

.

1.5 Invariance

1.5.1 Scaling

If B_t is Brownian motion then for every $r > 0$,

$$\{r^{-1/2} B_{tr}, t \geq 0\} \overset{\text{dist}}{=} \{B_t, t \geq 0\}.$$

This space-time scaling is known in the theory of parabolic PDE's.

1.5.2 Time Reversal

If B_t is Brownian motion then

$$\{B_{1-t} - B_1, 0 \leq t \leq 1\} \overset{\text{dist}}{=} \{B_t - B_0, 0 \leq t \leq 1\}.$$

1.5.3 Invariance of d-Dimensional Brownian Motion

The d-dimensional Brownian motion is invariant under isometries of the d-dimensional Euclidean space. It also inherits invariance properties of the 1-dimensional Brownian motion.

The following simple formula has profound meaning and far reaching consequences. Recall that if (B_t^1, B_t^2) is a two-dimensional Brownian motion then the density of B_1^j is $(1/\sqrt{2\pi}) \exp(-x_j^2/2)$ for $j = 1, 2$ and, therefore, the joint density of (B_1^1, B_1^2) is

$$\frac{1}{\sqrt{2\pi}} \exp(-x_1^2/2) \frac{1}{\sqrt{2\pi}} \exp(-x_2^2/2) = \frac{1}{2\pi} \exp(-(x_1^2 + x_2^2)/2)$$

$$= \frac{1}{2\pi} \exp(-|(x_1, x_2)|^2/2).$$

The normal distribution is the only distribution, in a sense, which has this property; see Sect. 1.10.

1.5.4 Conformal Invariance

Suppose that $D \subset \mathbb{C}$ is a planar domain and let $f : D \to \mathbb{C}$ be an analytic function. Suppose that B is a two-dimensional Brownian motion with $B_0 \in D$. Let $\tau_D = \inf\{t \geq 0 : B_t \notin D\}$ and

$$c(t) = \int_0^t |f'(B_s)|^2 ds, \qquad 0 \leq t < \tau_D. \tag{1.2}$$

Then $\{f(B_{c^{-1}(t)}), t \in [0, c(\tau_D))\}$ is a (stopped) two-dimensional Brownian motion.

1.6 Cameron-Martin-Girsanov Formula

Suppose that $f(0) = 0$ and $\int_0^1 f'(s)^2 ds < \infty$. Then the Radon-Nikodym derivative of the distribution of $\{B_t + f(t), t \geq 0\}$ with respect to the distribution of $\{B_t, t \geq 0\}$ is given by

$$\exp\left(\int_0^1 f'(s) dB_s - \frac{1}{2}\int_0^1 f'(s)^2 ds\right).$$

1.7 Brownian Motion and PDE's

The heat equation (normalized conveniently for probabilistic applications) is

$$\frac{\partial}{\partial t} u(x,t) = \frac{1}{2}\Delta_x u(x,t),$$

where $u : \mathbb{R} \times [0,\infty) \to \mathbb{R}$. The function $u(x,t)$ may be interpreted as the temperature at location x at time t.

Let $\mathbb{P}^x$ denote the distribution of Brownian motion starting from $B_0 = x$ and for a probability measure μ on $\mathbb{R}$, let $\mathbb{P}^\mu$ denote Brownian motion with the (initial) distribution of B_0 equal to μ. An analogous notation will be used for expectations.

The heat equation has the following "forward representation" in terms of Brownian motion. Let $\mu(dx) = u(x,0)dx$. Then for $t \geq 0$,

$$u(x,t)dx = \mathbb{P}^\mu(B_t \in dx).$$

The following "backward representation" is a form of the Feynman-Kac Formula (see Sect. 1.7.1). For $t \geq 0$ and $x \in \mathbb{R}$,

$$u(x, t) = \mathbb{E}^x u(B_t, 0).$$

1.7.1 Feynman-Kac Formula

Various authors apply the term "Feynman-Kac Formula" to various equations linking PDE's and probability. Quite often, the term refers to a model that involves a "potential."

Suppose that $D \subset \mathbb{R}^d$ is a bounded domain (open connected set) and B is a d-dimensional Brownian motion (not necessarily starting from 0). Let $\tau_D = \inf\{t \geq 0 : B_t \notin D\}$. Suppose that $V : \overline{D} \to \mathbb{R}$ and $f : \partial D \to \mathbb{R}$ are continuous functions. The function f represents "Dirichlet boundary conditions" and V is a "potential." Then the function

$$u(x) = \mathbb{E}^x \left(f(B_{\tau_D}) \exp \left(- \int_0^{\tau_D} V(B_s) ds \right) \right), \qquad x \in D,$$

satisfies the equation

$$\frac{1}{2} \Delta u(x) - V(x) u(x) = 0, \qquad x \in D,$$

with boundary values given by f. If $V \equiv 0$ then u is harmonic in D.

1.8 Mild Modifications of Brownian Motion

One could say that a process is a "mild modification" of Brownian motion if its generator is the Laplacian (with a suitable domain). Suppose that $D \subset \mathbb{R}^d$ is a domain. Three mild modifications of Brownian motion correspond to three classical problems in the theory of partial differential equations.

1. Brownian motion killed at the hitting time of ∂D corresponds to the Dirichlet problem.
2. Brownian motion reflected on ∂D corresponds to the Neumann problem.
3. Brownian motion absorbed on ∂D at a non-degenerate rate corresponds to the Robin problem.

1.9 Related Models

There are several well-studied classes of processes which are related to Brownian motion. Each of the following families of processes contains Brownian motion but it is much wider than any of the variants of Brownian motion mentioned so far. Each family can be used to model a range of natural or social phenomena for which Brownian motion would be a clearly inadequate model. Each family has a parameter or parameters that can be "calibrated" to fit the model to a real phenomenon. Only one value of the parameter corresponds to Brownian motion.

1.9.1 Diffusion Processes

One can define a diffusion process as a continuous Markov process. But it is common to think about a diffusion process as a solution to the stochastic differential equation (SDE) of the form

$$dX_t = \sigma(X_t)dB_t + \mu(X_t)dt.$$

More formally, we require that for $t \geq 0$,

$$X_t = X_0 + \int_0^t \sigma(X_s)dB_s + \int_0^t \mu(X_s)ds.$$

Thus defined diffusion processes are Markov. They are martingales only if $\mu \equiv 0$. Typically, they are not Gaussian but quite often their distributions have Gaussian tails.

1.9.2 Conditioned Brownian Motion

Let B denote multidimensional Brownian motion killed at the exit time τ_D from a domain D. Suppose that h is a positive harmonic function in D. The process X defined by

$$dX = dB + \frac{\nabla h}{h}(X_t)dt$$

is (a special case of) Doob's h-process. If the transition density function for B is denoted $p_t(x, y)$ then the transition density function for X is given by

$$p_t^h(x, y) = \frac{h(y)}{h(x)}p_t(x, y).$$

For a detailed treatment of h-processes and related topics, see [Bas95, Doo84, Dur84]. The book [Doo84] is a great reference source but it is hard to read.

1.9.3   Stable Processes

Heuristically speaking, the increments of Brownian motion B satisfy $dB \approx (dt)^{1/2}$. For any $\alpha \in (0, 2]$, there exists a Markov processes X with stationary independent increments (that is, a Lévy process) satisfying $dX \approx (dt)^{1/\alpha}$. Except for Brownian motion, these processes are discontinuous, non-Gaussian and their distributions have tails that are much "heavier" than Gaussian tails.

1.9.4   Fractional Brownian Motion

The singular term "fractional Brownian motion" refers to a family of processes X with the scaling property $dX \approx (dt)^{1/\alpha}$, same as for the stable processes. The similarities with stable processes end here. The range of the parameter is $1 < \alpha < \infty$. Every fractional Brownian motion is a continuous Gaussian process. Except for Brownian motion, these processes are not Markov and they do not have the martingale property.

1.10   Digression: Rotationally Invariant Distributions

If $d \geq 3$, the components of the d-dimensional vector $(X_1, X_2, \ldots, X_d)$ are independent, $\mathbb{P}((X_1, X_2, \ldots, X_d) = (0, \ldots 0)) = 0$ and $(X_1, X_2, \ldots, X_d)/|(X_1, X_2, \ldots, X_d)|$ has the uniform distribution on the unit sphere then $(X_1, X_2, \ldots, X_d)$ is Gaussian. See [Bry95, Theorem 4.2.3].

Chapter 2
Probabilistic Proofs of Classical Theorems

Probabilists noticed that some classical results in mathematical analysis can be interpreted using probability long time ago. The trend was started by Kakutani in mid-1940s (see [Kak44, Kak45a, Kak45b]) and followed by Burkholder [Bur76], Davis [Dav75, Dav79a] and Doob [Doo61], among others. It was pointed out by Burkholder [Bur76, p. 147] that probabilistic proofs were not necessarily shorter or easier than the analytic ones. They did provide new insight, though. This chapter is devoted to new probabilistic proofs of results previously proved using analytic techniques. Starting with Chap. 3, we will review results in analysis whose original (and, in many cases, unique, so far) proofs are probabilistic in nature.

2.1 Probabilistic Proofs of the Fundamental Theorem of Algebra

Theorem 2.1. *If $f : \mathbb{C} \to \mathbb{C}$, $f(z) = a_n z^n + a_{n-1} z^{n-1} + \ldots + a_1 z + a_0$ ($a_n \neq 0$, $a_0, \ldots, a_n \in \mathbb{C}$) is a polynomial of degree $n \geq 1$, then $f(\mathbb{C}) = \mathbb{C}$.*
  In particular, the equation $f(z) = 0$ has at least one root in $\mathbb{C}$.

I will present three probabilistic proofs of the fundamental theorem of algebra. The first proof, by Luo [Luo95], was published in Chinese and for this reason it is unlikely to be known to many mathematicians. I am grateful to Zhenqing Chen for translating [Luo95] to English (but I take responsibility for any mistakes in the presentation of the proof given below). Next I will present Zhenqing Chen's own improved version of Luo's proof (private communication). Finally, I will present a different proof invented independently by Pascu and published in [Pas05].

The proofs will use the following result on transience and recurrence. We say that a process X taking values in $\mathbb{R}^d$ is (neighborhood) *recurrent* if for every non-empty open set $U \subset \mathbb{R}^d$ and every $t > 0$, with probability 1, there exists $s > t$ such that $X_s \in U$. We say that a process X taking values in $\mathbb{R}^d$ is *transient* if $\lim_{t \to \infty} |X_t| = \infty$, a.s.

K. Burdzy, *Brownian Motion and its Applications to Mathematical Analysis*,
Lecture Notes in Mathematics 2106, DOI 10.1007/978-3-319-04394-4_2,
© Springer International Publishing Switzerland 2014

Proposition 2.2. *The d-dimensional Brownian motion is recurrent if and only if* $d \leq 2$.

See [MP10, Theorem 3.20] for the proof.

2.1.1 Luo's Proof

See [Luo95]. Consider a polynomial $f : \mathbb{C} \to \mathbb{C}$, that is, suppose that $f(z) = \sum_{k=0}^{n} a_k z^k$. Assume that $a_k \neq 0$ for some $k > 0$. It is easy to see that

$$\lim_{|z| \to \infty} |f(z)| = \infty. \tag{2.1}$$

We have $|f'(z_1)| > 0$ for some z_1 and, therefore, by the continuity of f', for some $r, \varepsilon > 0$ we have $|f'(z)| > \varepsilon$ for $z \in \mathcal{B}(z_1, r)$. Let B be Brownian motion with $B_0 = 0$. Brownian motion is neighborhood recurrent so B will return to $\mathcal{B}(z_1, r)$ infinitely often, a.s. This implies easily that, a.s.,

$$\int_0^\infty |f'(B_s)|^2 ds = \infty. \tag{2.2}$$

Let X_t be the time-changed process $f(B_t)$ as in Sect. 1.5.4, so that X is Brownian motion. Comparing (1.2) and (2.2), we see that X_t is defined for all $t > 0$, that is, X is not stopped.

Consider any $z \in \mathbb{C}$. By the recurrence of X, for every integer $n \geq 1$, a.s., X will visit $\mathcal{B}(z, 1/n)$, say, at a time $t_{z,n}$. If we write $w_n = B_{t_{z,n}}$ then $f(w_n) \in \mathcal{B}(z, 1/n)$. By compactness, there is a subsequence $\{w_{n_k}\}$ which either converges to a number $w_\infty \in \mathbb{C}$ or such that $\lim_{k \to \infty} |w_{n_k}| = \infty$. The last condition cannot hold because of (2.1) and the fact that $f(w_{n_k}) \in \mathcal{B}(z, 1/n_k)$. Hence, $\lim_{k \to \infty} w_{n_k} = w_\infty$. This and continuity of f imply that $f(w_\infty) = z$. This completes the proof.

2.1.2 Chen's Proof

The initial part of this proof is the same as Luo's. We copy it to make Chen's proof self-contained.

Consider a polynomial $f : \mathbb{C} \to \mathbb{C}$, that is, suppose that $f(z) = \sum_{k=0}^{n} a_k z^k$. Assume that $a_k \neq 0$ for some $k > 0$. It is easy to see that

$$\lim_{|z| \to \infty} |f(z)| = \infty. \tag{2.3}$$

We have $|f'(z_1)| > 0$ for some z_1 and, therefore, by the continuity of f', for some $r, \varepsilon > 0$ we have $|f'(z)| > \varepsilon$ for $z \in \mathcal{B}(z_1, r)$. Let B be Brownian motion with $B_0 = 0$. Brownian motion is neighborhood recurrent so B will return to $\mathcal{B}(z_1, r)$ infinitely often, a.s. This implies that, a.s.,

$$\int_0^\infty |f'(B_s)|^2 ds = \infty. \tag{2.4}$$

Let X_t be the time-changed process $f(B_t)$ as in Sect. 1.5.4, so that X is Brownian motion. Comparing (1.2) and (2.4), we see that X_t is defined for all $t > 0$, that is, X is not stopped.

It will suffice to show that there exists $w \in \mathbb{C}$ such that $f(w) = 0$. Suppose otherwise. Then (2.3), continuity of f and a compactness argument imply that $r := \inf\{|f(z)| : z \in \mathbb{C}\} > 0$. Hence, $\inf\{|X_t| : t \geq 0\} = \inf\{|f(B_t)| : t \geq 0\} \geq r > 0$. This contradicts neighborhood recurrence of two-dimensional Brownian motion. The proof is complete.

2.1.3 Pascu's Proof

See [Pas05]. We will need the following support theorem for Brownian motion (see [Bas95, p. 59] for a proof).

Theorem 2.3. *If $\varphi : [0, t] \to \mathbb{R}^d$ is continuous, $d \geq 1$, B_t is a d-dimensional Brownian motion starting at $B_0 = \varphi(0)$ and $\varepsilon > 0$, then*

$$\mathbb{P}^{\varphi(0)}\left(\sup_{s \leq t}\|B_s - \varphi(s)\| < \varepsilon\right) > 0. \tag{2.5}$$

Recall the "conformal invariance" of Brownian motion from Sect. 1.5.4.

Given a domain D, a closed curve $\gamma \subset D$ is said to be homotopic to zero in D if the curve γ can be deformed continuously in D to a constant curve. Homotopy is a topological property (it is preserved under continuous mappings).

Suppose that f is a polynomial and $z \in \mathbb{C}$ is not in the range of f. Then by the support theorem, a planar Brownian motion trajectory will wind around z with positive probability; if we choose this Brownian motion to be the image under f of a Brownian motion in the domain of f (that is, $\mathbb{C}$), this allows us to construct (on a set of positive probability) a closed curve γ in $\mathbb{C}$ which is homotopic to zero, while its image $\Gamma = f(\gamma)$ is not, thus obtaining a contradiction.

2.2 Privalov's Theorem

See [Bur76]. Consider a domain $D \subset \mathbb{C}$ with smooth boundary and suppose that $x \in \partial D$. Let $S(x, \alpha)$ denote the open cone with the vertex x, angle $2\alpha \in (0, \pi)$, the axis of symmetry containing the normal vector to ∂D at x, and "pointing inside" the domain, that is, for small $\varepsilon > 0$, $\mathcal{B}(x, \varepsilon) \cap S(x, \alpha) \subset D$. The set $S(x, \alpha)$ is called the Stolz angle in this context. We say that a function $f : D \to \mathbb{C}$ has a non-tangential limit y at $x \in \partial D$ if for every $\alpha \in (0, \pi/2)$,

$$\lim_{z \to x, z \in D \cap S(x, \alpha)} f(z) = y.$$

Recall the concept of conditioned Brownian motion (h-process) from Sect. 1.9.2. Let $K_x(y)$ be the Poisson kernel at $x \in \partial D$, that is $y \to K_x(y)$ is the unique, up to a multiplicative constant, non-zero positive harmonic function in D which vanishes continuously on $\partial D \setminus \{x\}$. If $h = K_x$ then the h-process X converges to x at its lifetime τ, a.s.

Let B be Brownian motion killed at the exit time from D. We define the harmonic measure on ∂D by $\mu_x(dz) := \mathbb{P}(B_{\tau-} \in dz \mid B_0 = x)$. The process B starting from $B_0 = x$ is a mixture of K_z-processes in the sense that if the distribution of K_z-process starting from x is denoted by $\mathcal{L}_x^z$ then the distribution of B evaluated at an event A is equal to $\int_{\partial D} \mathcal{L}_x^z(A) \mu(dz)$.

We say that a function $f : D \to \mathbb{C}$ has a minimal-fine limit y at x if, a.s.,

$$\lim_{t \uparrow \tau} f(X_t) = y, \tag{2.6}$$

where X is K_x-process.

Homework 2.4. (i) If f is a harmonic function in D and f has a minimal fine limit y at $x \in \partial D$ then its non-tangential limit at x exists and is equal to y. (See [Dav79b].)

(ii) Find a continuous function f in D such that it has a minimal fine limit y at $x \in \partial D$ and it does not have a non-tangential limit at x.

(iii) The event $\{\tau < \infty\}$ has probability 0 or 1.

(iv) If for some set A we have $\mathbb{P}(\lim_{t \uparrow \tau} \mathbf{1}_A(f(X_t))) > 0$ then the probability is equal to 1.

Theorem 2.5. *If an analytic function in the unit disc has the minimal-fine limit equal to 0 almost everywhere on the boundary of the disc then it is identically equal to 0.*

Sketch of the Proof. Suppose that f is an analytic function in the unit disc, it has the minimal-fine limit equal to 0 almost everywhere on the boundary of the disc and it is not identically equal to 0. Suppose that z is a point in the unit disc such that $f(z) \neq 0$. Let B be Brownian motion with $B_0 = z$, killed at the exit time from the disc and let X be the time change of $f(B)$ such that X is a (stopped) Brownian motion (see Sect. 1.5.4). Let τ denote the lifetime of X (it could be finite or infinite). Since f has the minimal-fine limit equal to 0 almost everywhere on the boundary of the disc, we see that $\lim_{t \uparrow \tau} X_t = 0$. This contradicts the fact that Brownian motion does not hit 0, a.s. (since it starts from $f(z) \neq 0$) and it does not converge to 0 as $t \to \infty$. $\qquad \square$

2.3 Plessner's Theorem

See [Doo61, Dav79a].

Theorem 2.6. *Suppose that f is an analytic function in the unit disc D. Then for almost every $x \in \partial D$, either*

(i) *f has a minimal fine limit at x, or*
(ii) *the minimal fine cluster set at x is $\mathbb{C}$.*

We can rephrase (ii) as follows. For every open set $U \subset \mathbb{C}$, the function $\mathbf{1}_{f^{-1}(U)}$ does not have the minimal fine limit 0 at x.

Sketch of the Proof. Suppose that f is an analytic function in the unit disc. Let B be Brownian motion with $B_0 = 0$, killed at the exit time from the disc and let X be the time change of $f(B)$ such that X is a (stopped) Brownian motion (see Sect. 1.5.4). Let τ denote the lifetime of X (it could be finite or infinite). Recall that B is a mixture of K_z-processes for $z \in \partial D$ and see Homework 2.4 (iii)–(iv). If, for a given $z \in \partial D$, we have $\tau < \infty$ then f has a minimal fine limit at x. If $\tau = \infty$, a.s., then X visits every neighborhood in $\mathbb{C}$ infinitely often and, therefore, (ii) holds. □

2.4 Picard's Theorem

The probabilistic proof of Picard's theorem is an illustration of the fact that probabilistic methods are sometimes much harder and less intuitive than the analytic methods used to prove the same result.

A function $f : \mathbb{C} \to \mathbb{C}$ is called entire if it is analytic in the whole complex plane $\mathbb{C}$.

Theorem 2.7 (Picard). *Suppose that $f : \mathbb{C} \to \mathbb{C}$ is a non-constant entire function. Then $\mathbb{C} \setminus f(\mathbb{C})$ contains at most one point.*

Proof. (Sketch) Suppose that f is entire and $\mathbb{C} \setminus f(\mathbb{C})$ contains at least two points. Composing f with a linear function, we may suppose that the range of f does not contain -1 and 1. Suppose that B_t is a two-dimensional Brownian motion, $B_0 = 0$, and let X_t be the time-changed process $f(B_t)$ so that X_t is also a planar Brownian motion. For any $\varepsilon > 0$, the process B_t will return to the ball $\mathcal{B}(0, \varepsilon)$ infinitely often, a.s. Choose $\varepsilon > 0$ so small that $-1, 1 \notin f(\mathcal{B}(0, \varepsilon))$. Every closed loop in $\mathbb{C}$ is homotopic to a single point. In particular, if $B_t \in \mathcal{B}(0, \varepsilon)$ then the closed loop L consisting of $\{B_s, s \in [0, t]\}$ and the line segment I between B_0 and B_t is homotopic to 0. Since f is continuous, it follows that the closed loop consisting of $\{f(B_s), s \in [0, t]\}$ and $f(I)$ is homotopic to 0. However, we make the following crucial claim, which will be discussed after the proof.

Brownian motion X_t entangles itself around -1 and 1 as $t \to \infty$, a.s. (2.7)

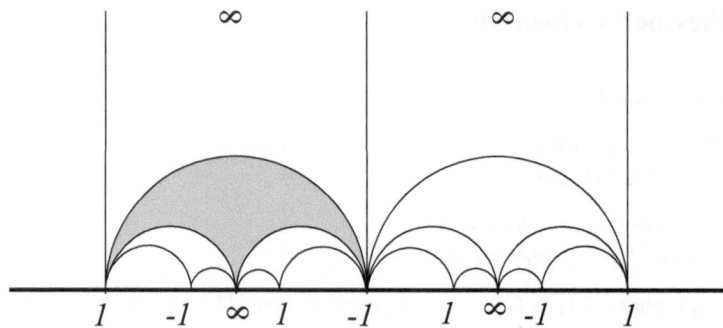

Fig. 2.1 The upper half-plane is the conformal image of the Riemann surface D_1 corresponding to the plane punctured at -1 and 1. The *shaded area* represents the image of a half-plane, a part of a sheet of the Riemann surface

In other words, with probability 1, for sufficiently large random time T, and all $t \geq T$, no closed loop consisting of $\{f(B_s), s \in [0, t]\}$ and a continuous path joining $f(B_0)$ and $f(B_t)$ inside $f(\mathcal{B}(0, \varepsilon))$ is homotopic to 0. This contradiction completes the proof.   □

The first proof of claim (2.7) was given in [IM74, Sect. 7.18]. Here is a sketch. Consider $\mathbb{C} \setminus \{-1, 1\}$ and cut it along the following line segments on the real axis: $(-\infty, -1), (-1, 1)$ and $(1, \infty)$. This leads naturally to a Riemann surface (let's call it D_1). The process X_t may be considered to be a Brownian motion on D_1 instead of $\mathbb{C} \setminus \{-1, 1\}$. Let $g : D_1 \to H$ be a conformal mapping of D_1 onto the upper half-plane H, illustrated in Fig. 2.1. A time change Y_t of $g(X_t)$ is a Brownian motion in H stopped at the hitting time of the real axis. Since the process Y_t will eventually exit the image of any given half-plane (such as the shaded area), the process X_t must get entangled around -1 and 1. The fully rigorous version of the proof requires the "monodromy theorem" (see [Ahl78, Sect. 8.1.6]) and "modular functions." The above partly probabilistic proof is not much different in spirit from the core argument in the purely analytic proof of Picard's theorem given in [Ahl78, Sect. 8.3.1].

The first fully probabilistic proof of the claim (2.7) appeared in [Dav75]. It was followed by two different versions presented in [Bas95, Sect. V.2, p. 320] and [Dur84, Sect. 5.6] (it seems that both of these versions contain gaps; see errata for [Bas95]). I do not find any of the probabilistic proofs of (2.7) any more attractive than the proof given in [IM74].

Probability is needed if one wants to prove results about windings stronger than (2.7). Here is an illustration. Suppose that B_t is a planar Brownian motion with $B_0 = 0$. Let $\theta_-(t) = \arg(B_t + 1)$, where arg is chosen so that $\theta_-(0) = 0$ and $t \to \theta_-(t)$ is continuous a.s. We define $\theta_+(t) = \arg(B_t - 1)$ in the analogous way. It was proved in [Mou88] that, a.s.,

$$\lim_{t \to \infty} \theta_-^2(t) + \theta_+^2(t) = \infty. \tag{2.8}$$

Note that (2.8) says more than that a typical Brownian path will a.s. become entangled around -1 and 1 in the punctured plane (i.e., the plane with -1 and 1 deleted). The claim (2.8) says that for sufficiently large t_0, the Brownian path will be entangled either around -1 or 1 at every $t \geq t_0$. See [PY86, Sect. 7] and [Gru98] for various advanced results on windings.

2.5 McMillan's Twist Point Theorem

This section is based on [Bur90]. McMillan's twist point theorem is a challenge for probabilists. A number of people feel that the theorem has an, as yet undiscovered, proof based mostly on stochastic analysis. The most convincing application of probability to this theorem seems to be the rigorous statement of a conjecture describing a generalization of McMillan's theorem to higher dimensions (see Problem 2.15).

Let $D_* = \{x \in \mathbb{C} : \operatorname{Re} x > 0\}$. Recall the definition of a Stolz angle $S(x, \alpha)$ from Sect. 2.2. In this section, for any $D \subset \mathbb{C}$, we will consider only functions $f : D \to D_1$ which are analytic, one-to-one and onto D_1.

The Green function of a Greenian set $D \subset \mathbb{C}$ will be denoted G_D. The function $K_D^x(y, z) := G_D(y, z)/G_D(y, x)$ is called the Martin function (for $y \neq x$). There exists a unique up to a homeomorphism compactification D^M of D such that K_D^x may be extended continuously to $D^M \setminus \{x\} \times D$. It will be called the Martin compactification and $\partial^M D := D^M \setminus D$ will be called the Martin boundary of D. For simply connected planar domains, the Martin boundary $\partial^M D$, the minimal Martin boundary $\partial_1^M D$ and the Caratheodory prime end boundary coincide (see [Doo84, Pom75] for definitions).

If $A \subset D$, $y \in \partial_1^M D$ and $\lim_{z \in A, z \to y} f(z)/K_D^x(y, z) = \infty$ for some superharmonic function f in D then we say that the set A is minimal thin in D at y. To define minimal fine topology, call a set $A \subset D$ a minimal fine neighborhood of $y \in \partial_1^M D$ if $D \setminus A$ is minimal thin in D at x. The limits in the minimal fine topology will be denoted mf–lim. For a probabilistic interpretation of the minimal fine topology, see (2.6).

The Green function, minimal thinness and minimal fine topology are conformal invariants.

Lemma 2.8. *(i) A simply connected open set $D \subset D_*$ is a minimal fine neighborhood of 0 in D_* if and only if there exist $\epsilon > 0$ and a nonnegative Lipschitz function $h : \mathbb{R} \to \mathbb{R}$ with constant 1 such that*

$$\int_{-1}^{1} h(r) r^{-2} dr < \infty \qquad and \qquad \{x \in \mathbb{C} : \operatorname{Re} x > h(\operatorname{Im} x), |x| < \epsilon\} \subset D.$$

(ii) If D is a simply connected open set and a minimal fine neighborhood of 0 in D_* then for each $\alpha < \pi/2$ there exists $\epsilon > 0$ such that $\{x \in S(0, \alpha) : |x| < \epsilon\} \subset D$.

We will say that a function f defined on D_* has the angular limit a at $x \in \partial D_*$ if $\lim_{z \in S(x,\alpha), z \to x} f(z) = a$ for every $\alpha < \pi/2$.

Lemma 2.9 (See [Dav79b]). *Suppose that f is analytic in D_* and $z \in \partial D_*$. Suppose that mf–$\lim_{x \to z} f(x) = a$. Then the angular limit of f at z exists and equals a.*

Suppose that $f : D_* \to D$ and $y \in \partial D_*$. If the limit of $f(x)$ exists for $x \to y$ along a single continuous line in D_* then it exists in the minimal fine topology. By Lemma 2.9, it exists as an angular limit as well. The common value of the limits will be denoted $f(y)$, provided they exist. If $0 \in \partial D$ and $\{x \in \mathbb{R} : 0 < x < \epsilon\} \subset D$ for some $\epsilon > 0$ then this line segment defines a prime end in D which will be denoted 0^M.

Definition 2.10. For a function $f : D_* \to D$, the minimal fine limit of f' at $x \in \partial D_*$ will be called the minimal fine derivative of f at x and denoted mf–$f'(x)$, provided it exists. The angular limit of f' at $x \in \partial D_*$ will be called the angular derivative of f at x and denoted a–$f'(x)$, provided it exists.

Proposition 2.11. *If mf–$f'(x) = a$ then a–$f'(x) = a$.*

The converse statement is false.

Theorem 2.12. *If $f : D_* \to D$, $x \in \partial D_*$ and mf–$f'(x) = a \in \mathbb{C}$ then $f(x)$ exists and*

$$\text{mf–}\lim_{z \in D_*, z \to x} \frac{f(z) - f(x)}{z - x} = a$$

.

Definition 2.13. Consider a function $f : D_* \to D$. A point $x \in \partial D_*$ will be called an f-twist point if $f(x)$ exists and for every continuous path $\{\Gamma(t), t \in (0, 1)\}$ in D_* with $\lim_{t \to 0} \Gamma(t) = x$, one has

$$-\liminf_{t \to 0} \arg(f(\Gamma(t)) - f(x)) = \limsup_{t \to 0} \arg(f(\Gamma(t)) - f(x)) = \infty,$$

where arg is chosen so that $t \to \arg(f(\Gamma(t)) - f(x))$ is continuous.

The following result is a slight generalization of the well known McMillan's twist point theorem [McM69].

Theorem 2.14. *Suppose that $f : D_* \to D$. Then for almost all $x \in \partial D_*$, either mf–$f'(x)$ exists and belongs to $\mathbb{C} \setminus \{0\}$ or x is an f-twist point.*

There exists a proof of McMillan's theorem based partially on stochastic analysis; see the articles by O'Neill [O'N11, O'N12]. O'Neill's project and a purely analytic approach in [ACTDBP06] are attempts to provide a proof for the twist point theorem which is not based on complex analysis, with view towards multidimensional generalizations of the theorem.

Let $d \geq 2$, $D \subset \mathbb{R}^d$ be a bounded open set, and B_t be d-dimensional Brownian motion starting from $B_0 \in D$. Let $\tau = \inf\{t > 0 : B_t \notin D\}$ be the exit time from D. Let $A = A(D, \omega)$ be the set of "asymptotic directions of approach," defined as the set of all cluster points of

$$\frac{B_t - B_\tau}{|B_t - B_\tau|},$$

as $t \uparrow \tau$. If D is a simply connected planar domain then, by Theorems 2.12 and 2.14, with probability 1, A is either a semicircle or circle. In some domains D, the set A is a circle with a non-trivial probability, i.e., with a probability strictly between 0 and 1.

Problem 2.15. Is it true that for every $d \geq 2$ and every d-dimensional open set D, the set A of asymptotic directions of approach is either a sphere or a hemisphere, a.s.?

Chapter 3
Overview of the "Hot Spots" Problem

The remaining part of these lecture notes is concerned with "modern" (post-1990) research at the intersection of probability and analysis. We start with the "hot spots" problem, for several reasons. The "hot spots" conjecture is quite easy to state and to understand. Yet, to this day, the conjecture has been proved only for a very limited family of domains. Hence, it has a great potential as a source of interesting problems (related questions) and as a testing ground for new techniques.

In 1974 Jeff Rauch stated a problem at a conference, since then referred to as the "hot spots conjecture" (the conjecture was not published in print until 1985, in a book by Kawohl [Kaw85]). Informally speaking, the conjecture says that the second Neumann eigenfunction for the Laplacian in a Euclidean domain attains its maximum and minimum on the boundary. There was hardly any progress on the conjecture for 25 years but a number of papers have been published in recent years, on the conjecture itself and on problems related to or inspired by the conjecture. This chapter will review some of this body of research and techniques used in it, with focus on author's own research and probabilistic methods used in proofs of analytic results. First, we will state and explain the conjecture. Then we will review some key results on the conjecture and related problems. This will be followed by a review of some techniques used in the proofs.

In order to explain the intuitive contents of the hot spots conjecture, we will start with the heat equation. Suppose that D is an open connected bounded subset of $\mathbb{R}^d$, $d \geq 1$, with smooth boundary. Let $u(t, x), t \geq 0, x \in D$, be the solution of the heat equation $\partial u / \partial t = \frac{1}{2} \Delta_x u$ in D with the Neumann boundary conditions and the initial condition $u(0, x) = u_0(x)$. That is, $u(t, x)$ is a solution to the following initial-boundary value problem,

$$\begin{cases} \frac{\partial u}{\partial t}(t, x) = \frac{1}{2}\Delta_x u(t, x), & x \in D, t > 0, \\ \frac{\partial u}{\partial n}(t, x) = 0, & x \in \partial D, t > 0, \\ u(0, x) = u_0(x), & x \in D, \end{cases} \tag{3.1}$$

K. Burdzy, *Brownian Motion and its Applications to Mathematical Analysis*, Lecture Notes in Mathematics 2106, DOI 10.1007/978-3-319-04394-4_3, © Springer International Publishing Switzerland 2014

where $\mathbf{n}(x)$ denotes the inward normal vector at $x \in \partial D$. The problem can be generalized to less smooth domains. For example, in Lipschitz domains, the Neumann boundary condition, $\frac{\partial u}{\partial \mathbf{n}}(t, x) = 0$, would be required to hold not for all but for almost all (with respect to the surface area measure) $x \in \partial D$. The long time behavior of a "generic" solution (i.e., the solution corresponding to a "typical" initial condition) can be derived from the properties of the second eigenfunction using the following eigenfunction expansion. Under suitable conditions on the domain, such as convexity or Lipschitz boundary, and for a "typical" initial condition $u_0(x)$, we have

$$u(t, x) = c_1 + c_2 \varphi_2(x) e^{-\mu_2 t} + R(t, x), \tag{3.2}$$

where $c_1 \in \mathbb{R}$ and $c_2 \neq 0$ are constants depending on the initial condition, $\mu_2 > 0$ is the second eigenvalue for the Neumann problem in D, $\varphi_2(x)$ is a corresponding eigenfunction, and $R(t, x)$ goes to 0 faster than $e^{-\mu_2 t}$, as $t \to \infty$. Note that the first eigenvalue is equal to 0 and the first eigenfunction is constant. Suppose that $\varphi_2(x)$ attains its maximum at the boundary of D. Under this assumption, for "most" initial conditions $u_0(x)$, if z_t is a point at which the function $x \to u(t, x)$ attains its maximum, then the distance from z_t to the boundary of D tends to zero as $t \to \infty$. In other words, the "hot spots" move towards the boundary.

Conjecture 3.1 (Hot Spots Conjecture). The second eigenfunction for the Laplacian with Neumann boundary conditions in a bounded Euclidean domain attains its maximum at the boundary.

The above version of the hot spots conjecture is somewhat ambiguous as it does not specify whether the maximum has to be strict, i.e., whether the eigenfunction can attain the same maximal value somewhere in the interior of the domain; it does not address the question of what might happen when the second eigenvalue is not simple, i.e., whether all eigenfunctions corresponding to the second eigenvalue have to satisfy the conjecture (in some domains, for example, the square, there are infinitely many eigenfunctions corresponding to the second eigenvalue). If φ_2 is an eigenfunction corresponding to μ_2 then so is $-\varphi_2$. The maximum of φ_2 is the minimum of $-\varphi_2$. The conjecture does not state whether both the maximum and the minimum have to be attained at the boundary. It turns out that a precise statement of the conjecture is not needed because the results do not depend in a subtle way on its formulation.

The hot spots conjecture can be justified by appealing to our physical intuition and by examples amenable to explicit analysis. Heuristically speaking, the "heat" and "cold" are substances that annihilate each other so it is easy to believe that the hottest and coldest spots lie as far as possible from each other, hence on the boundary of the domain. One can find explicit formulas for the eigenfunctions in some simple domains, for example, in a rectangle $[0, a] \times [0, b]$ with $a > b > 0$, we have $\varphi_2(x_1, x_2) = \cos(\pi x_1 / a)$. All such explicit examples support the hot spots conjecture, i.e., the second eigenfunction attains the maximum on ∂D in simple domains such as rectangles and balls.

Fig. 3.1 An example of a lip domain

3.1 Main Theorems on the "Hot Spots" Problem

For 25 years, from 1974 to 1999, almost nothing was known about the "hot spots" conjecture. A notable exception was a result by Kawohl that appeared in his book [Kaw85] in 1985. Kawohl proved that if a set $D \subset \mathbb{R}^d$ is a cylindrical domain, i.e., if $d > 1$, and D can be represented as $D = D_1 \times [0, 1]$ for some $D_1 \subset \mathbb{R}^{d-1}$, then the hot spots conjecture holds for D. This result has a simple proof based on the factorization of eigenfunctions in cylindrical domains. Kawohl's most lasting contributions are the realization that one should restrict attention to some classes of domains, and the statement of the currently most significant open problem in the area—Kawohl suggested that the hot spot conjecture might not be true in general but it should be true for convex domains.

The next publication on the hot spots conjecture, [BB99], appeared in 1999. The paper contained the proof of the hot spots conjecture for two classes of planar domains: domains with a line of symmetry and "lip" domains, to be described shortly. The results were not complete, in the sense that the authors imposed some extra "technical" assumptions on domains in each family. Most of these extra assumptions were removed for symmetric domains by Pascu [Pas02] and for "lip" domains in [AB04].

A "lip" domain is a bounded planar domain such that its boundary consists of two graphs of Lipschitz functions with the Lipschitz constant equal to 1 (see Fig. 3.1). For example, any obtuse triangle (i.e., a triangle with an angle greater than $\pi/2$) is a lip domain if it is properly oriented (see Fig. 8.1 in Chap. 8).

Theorem 3.2. *The hot spots conjecture holds for $D \subset \mathbb{R}^2$ if*

(i) [BB99, Pas02] D is convex and has a line of symmetry, or
(ii) [BB99, AB04] D is a lip domain.

The methods and techniques developed in [BB99] to prove the hot spots conjecture for some classes of domains turned out to be useful also in deriving negative results. The first of such results appeared in [BW99] in 1999. The authors of [BW99] showed that there exists a planar domain where the second eigenvalue is simple and the eigenfunction corresponding to the second eigenvalue attains its maximum in the interior of the domain. This result was strengthened in [BB00], where it was shown that in some other planar domain, the second eigenvalue is simple and the second eigenfunction attains both its minimum and maximum in the interior of the domain. The domain constructed in [BB00] had many holes and the one constructed in [BW99] had two holes. The intuitive idea behind the examples constructed in [BW99] and [BB00] suggested that every counterexample to the hot spots conjecture in the plane must have at least two holes, and every counterexample in $\mathbb{R}^d$, $d \geq 3$, must have at least d handles. This turned out not to be true—another counterexample [Bur05] showed that there exists a planar domain with one hole and simple second eigenvalue, and such that the second eigenfunction attains both its maximum and minimum in the interior of the domain. The shape of the domain is much simpler than that of examples in [BW99] and [BB00]; see Fig. 4.2 in Sect. 4.1. We summarize these remarks as a theorem.

Theorem 3.3 ([BW99, BB00, Bur05]). *The hot spots conjecture fails for some domains $D \subset \mathbb{R}^2$.*

Much more recently, a new technique was used in [Miy09] to push the positive result in a new direction.

Theorem 3.4 ([Miy09]). *The hot spots conjecture holds for $D \subset \mathbb{R}^2$ if D is convex and* $\mathrm{diam}(D)^2/|D| < 1.378$.

When D is a disk, $\mathrm{diam}(D)^2/|D| \approx 1.273$. Hence, the condition in Theorem 3.4 indicates that D is a nearly circular planar convex domain. However, no symmetry of the domain is assumed.

Since Theorem 3.2 (ii) shows that the hot spots conjecture holds for (some) long and thin domains and Theorem 3.4 shows that the conjecture holds for convex and nearly circular domains, one can reasonably expect that the conjecture is true for the domains that lie "between" these two extremes. The following are, in author's opinion, the most significant open problems in this area. The first one was proposed by Kawohl in [Kaw85], and the second one is well known among researchers interested in the subject.

Problem 3.5. (i) [Kaw85] Does the hot spots conjecture hold for bounded convex domains $D \subset \mathbb{R}^d$ for all $d \geq 1$?

(ii) Does the hot spots conjecture hold for bounded simply connected planar domains?

As a sign of changing times, a "social network" project [Pol] has been created. The project is devoted to the following particular case of the hot spots conjecture.

Problem 3.6. Does the hot spots conjecture hold for all triangles?

Atar investigated in [Ata01] a class of multidimensional domains—this seems to be the only paper (except for an early result in [Kaw85]) that contains results on the multidimensional version of the problem.

It was known for a long time, as a "folklore" among experts in the field, that the hot spots conjecture does not hold for manifolds, see, e.g., remarks to this effect in [BB99] or [BB00]. The first rigorous paper studying the hot spots problem for manifolds was published by Freitas [Fre02].

3.2   Results Related to the "Hot Spots" Problem

The hot spots conjecture inspired a number of papers on the properties of Neumann eigenfunctions. We will review those that seem to be the closest in spirit to the original conjecture. For a review of research in related areas, see [NTY01].

First of all, we mention a paper by Hempel et al. [HSS91], which had appeared in 1991, long time before the current interest in the hot spots conjecture developed. The authors studied the spectrum of the Neumann Laplacian in bounded Euclidean domains with non-smooth boundaries. Roughly speaking, their results show that the spectrum does not need to be discrete; in a sense, it can be completely arbitrary. For this reason, the hot spots conjecture must be limited to domains where the spectrum is discrete, such as domains with Lipschitz boundaries.

Athreya [Ath00] showed that some monotonicity properties of Neumann eigenfunctions hold also for solutions of some semi-linear partial differential equations related to a class of stochastic processes known as "superprocesses." He adapted probabilistic techniques used in research on the hot spots conjecture to the new setting.

Jerison [Jer00] found the location (in an asymptotic sense) of the nodal line (i.e., the line where the eigenfunction vanishes) of the second Neumann eigenfunction in long and thin domains. The information about the location of the nodal line can be effectively used in research on the hot spots conjecture. This was first done in [BB99], where the nodal line was identified with the line of symmetry in domains possessing such a line. The knowledge of the nodal line can be used to transform the Neumann problem to a problem with mixed Neumann and Dirichlet conditions. In many cases, the mixed problem is much easier than the original one. Jerison and Nadirashvili considered in [JN00] convex planar domains with two perpendicular lines of symmetry, and showed that under these strong assumptions one can provide accurate information about the second eigenfunction. The location of the nodal line for the second eigenfunction was treated as a problem of its own interest in [AB02], where probabilistic techniques are used to give some results in this direction.

Although the paper by Ishige and Mizoguchi [IM03] is not devoted to the hot spots problem in the sense of this article, it is related because it studies geometric properties of the heat equation solutions.

Two papers by Bañuelos and Pang, one of them joint with Pascu [BP04, BPP04], are devoted to variations of the hot spots problem. The purpose of [BP04] is to prove an inequality for the distribution of integrals of potentials in the unit disk composed with Brownian motion which, with the help of Lévy's conformal invariance, gives another proof of Pascu's result [Pas02]. The paper [BPP04] investigates the "hot spots" property for the survival time probability of Brownian motion with killing and reflection in a planar convex domain whose boundary consists of two curves, one of which is an arc of a circle, intersecting at acute angles. This leads to the "hot spots" property for the mixed Dirichlet-Neumann eigenvalue problem in the domain with Neumann boundary conditions on one of the curves and Dirichlet boundary conditions on the other.

The monotonicity property of the Neumann heat kernel in the ball along the radius, a long standing conjecture, has been proved recently in [PG11] using probabilistic techniques.

3.3   Review of Selected Probabilistic Techniques

This section will focus on "essential probabilistic techniques," i.e., those techniques that involve stochastic processes and cannot be easily translated into the language of analysis. We will mainly discuss a probabilistic technique called "coupling." The technique was invented by Doeblin in 1930s. One can find a general review of this method in books by Lindvall [Lin92] and Chen [Che92a]. The most frequent application of the coupling technique consists of a construction of two processes on the same probability space, run with the same clock. Almost always, the processes are *not* independent and they meet at a certain time, called the coupling time. One usually tries to find a coupling with as small coupling time as possible. A distinguishing feature of applications of couplings in the context of the hot spots conjecture is that the properties of the coupling time usually do not matter, to the point that the coupling time is infinite for some of the couplings. Couplings were used for the first time to study the hot spots conjecture in [BB99] but that paper owes a lot to an earlier project, [BK00], devoted to a seemingly unrelated problem.

Many proofs of results on the hot spots conjecture are based on the eigenfunction expansion (3.2). First, a geometric property is proved for the heat equation solution and then it is translated into a statement about the second eigenfunction using (3.2), as $t \to \infty$.

Let X_t and Y_t be reflected Brownian motions in D starting from $x \in D$ and $y \in D$, resp. Then we can represent the solution $u(t, x)$ of the heat equation (3.1) as $u(t, x) = \mathbb{E}\, u_0(X_t)$, and similarly $u(t, y) = \mathbb{E}\, u_0(Y_t)$. We have by (3.2),

$$\varphi_2(x) - \varphi_2(y) = c_3 e^{\mu_2 t}(u(t, x) - u(t, y)) + R_1(t, x, y) \tag{3.3}$$
$$= c_3 e^{\mu_2 t}(\mathbb{E}\, u_0(X_t) - \mathbb{E}\, u_0(Y_t)) + R_1(t, x, y),$$

where $R_1(t, x, y)$ goes to 0 as $t \to \infty$. Without loss of generality we will assume that $c_3 > 0$. Suppose that we can prove for some initial condition u_0 that for all $t > 0$,

$$\mathbb{E}\, u_0(X_t) - \mathbb{E}\, u_0(Y_t) \le 0. \tag{3.4}$$

This and (3.3) will then show that $\varphi_2(x) \le \varphi_2(y)$. If the last inequality can be proved for an appropriate family of pairs (x, y), the hot spots conjecture will follow. We will next present a technique of proving (3.4).

For $x, y \in \mathbb{R}^2$, write $x \le y$ if the angle between $y - x$ and the positive horizontal half-line is within $[-\pi/4, \pi/4]$. Recall that a "lip" domain is a bounded planar domain such that its boundary consists of two graphs of Lipschitz functions with the Lipschitz constant equal to 1. Suppose that D is a lip domain and $x, y \in D$, $x \leq y$. Suppose that X_t and Y_t are reflected Brownian motions in D, driven by the same Brownian motion, and starting from x and y, resp. In other words,

$$X_t = x + B_t + \int_0^t \mathbf{n}(X_s)dL_s^X, \tag{3.5}$$

$$Y_t = y + B_t + \int_0^t \mathbf{n}(Y_s)dL_s^Y,$$

where $\mathbf{n}(z)$ is the unit inward normal vector at $z \in \partial D$ and L_s^X is the local time of X on the boundary of D, i.e., L^X is a non-decreasing process that does not increase when X is inside D. In other words,

$$\int_0^\infty \mathbf{1}_D(X_s)dL_s^X = 0.$$

Similar remarks apply to the formula for Y_t. For domains which are piecewise C^2-smooth, the existence of processes satisfying (3.5) follows from results of Lions and Sznitman [LS84]. For lip domains, one can use a result from [BBC05]. The existence of a strong unique solution to an equation analogous to (3.5) but in a multidimensional Lipschitz domain remains an open problem at this time although results in [BB08] strongly suggest that pathwise uniqueness fails in some Lipschitz domains.

We have assumed that the domain D is a lip domain so if the normal vector $\mathbf{n}(z)$ is well defined at $z \in \partial D$ (this is the case for almost all boundary points), it has to form an angle less than $\pi/4$ with the vertical. Then easy geometry shows that the "local time push" in (3.5), i.e., the term represented by the integral, is such that if $x \leq y$ then

$$X_t \leq Y_t \qquad \text{for all } t \geq 0. \tag{3.6}$$

Now consider a set $A \subset D$, such that both A and $D \setminus A$ have non-empty interiors and $\partial A \cap \partial(D \setminus A)$ is a vertical line segment. Suppose that A lies to the right of $D \setminus A$ and let the initial condition be $u_0(z) = \mathbf{1}_A(z)$. If (3.6) is satisfied, then for any fixed time $t \geq 0$, we may have $X_t, Y_t \in A$, or $X_t, Y_t \in D \setminus A$, or $X_t \in D \setminus A, Y_t \in A$, but we will never have $X_t \in A, Y_t \in D \setminus A$. This and the definition of u_0 imply (3.4). We combine this with (3.3) to conclude that $\varphi_2(x) \leq \varphi_2(y)$ for $x \leq y$. Any lip domain has the "leftmost" and "rightmost" points in the sense of the partial order "$\leq$" so our argument has shown that the maximum and the minimum of the second eigenfunction are attained at these two points. Hence, the hot spots conjecture holds in lip domains.

Planar domains with a line of symmetry have to be approached in a different manner. Suppose that $D \subset \mathbb{R}^2$ is symmetric with respect to a vertical line K and let

D_1 be the part of D lying to the right of K. Under some extra assumptions, the second eigenfunction φ_2 in D with the Neumann boundary conditions is antisymmetric with respect to K (this follows from a simple symmetrization argument). Therefore, φ_2 must vanish on K and we see that φ_2 is the first eigenfunction for the Laplacian in D_1 with the Neumann boundary conditions on $\partial D_1 \setminus K$ and Dirichlet boundary conditions on K. Such boundary conditions correspond to the Brownian motion in D_1 that is reflected on $\partial D_1 \setminus K$ and killed on K.

We will choose the initial condition u_0 to be identically equal to 1 in D_1. Let T_K^X be the hitting time of K by X and let T_K^Y have the analogous meaning for Y. The strategy is to construct Brownian motions X_t and Y_t in D_1, reflected on $\partial D_1 \setminus K$, killed on K, starting from x and y, and such that (3.4) holds not for a fixed time t but for an appropriate stopping time T. Let $T = T_K^X$. If we can show that X must hit K before Y does, then (3.4) follows and we have $\varphi_2(x) \leq \varphi_2(y)$ for this particular pair (x, y). See Sect. 5.5.2 for details of how to choose x and y and what assumptions one must make about the geometry of D to carry out the argument outlined above. Here we will describe a coupling of reflected Brownian motions (the "mirror" coupling) that keeps the two Brownian particles in a relative position that ensures that $T_K^X \leq T_K^Y$.

Let us start by defining the mirror coupling for free Brownian motions in $\mathbb{R}^2$. Suppose that $x, y \in \mathbb{R}^2$, $x \neq y$, and that x and y are symmetric with respect to a line M. Let X_t be a Brownian motion starting from x and let τ be the first time t with $X_t \in M$. Then we let Y_t be the mirror image of X_t with respect to M for $t \leq \tau$, and we let $Y_t = X_t$ for $t > \tau$. The process Y_t is a Brownian motion starting from y. The pair (X_t, Y_t) is a "mirror coupling" of Brownian motions in $\mathbb{R}^2$.

Next we turn to the mirror coupling of reflected Brownian motions in a half-plane $\mathcal{H}$, starting from $x, y \in \mathcal{H}$. One can construct reflected Brownian motions X_t and Y_t in $\mathcal{H}$, starting from x and y, so that they have the following properties. The processes X_t and Y_t behave like free Brownian motions coupled by the mirror coupling as long as they are both strictly inside $\mathcal{H}$. When one of the processes hits the boundary, the two particles cannot behave as a "free" mirror coupling in the whole plane. Let M be the line of symmetry for x and y and $H = M \cap \partial \mathcal{H}$. Then for every t, the distance from X_t to H is the same as for Y_t. Let M_t be the line of symmetry for X_t and Y_t. The "mirror" M_t may move, but only in a continuous way, while the point $M_t \cap \partial \mathcal{H} = H$ will never move. The absolute value of the angle between the mirror and the normal vector to $\partial \mathcal{H}$ at H can only decrease. The processes stay together after the first time they meet. The most important property of the mirror coupling is that the two processes X_t and Y_t remain at the same distance from a fixed point, the "hinge" H.

Since the reflecting particle cannot sense the global shape of the domain, the above description of the mirror coupling in a half-plane can be applied to describe the possible motions of the mirror (the line of symmetry between the processes) in a polygonal domain whenever only one of the processes is on the boundary. This simple recipe breaks down when the two processes hit the boundary at the same time. It is not obvious that two processes forming a mirror coupling can indeed hit the boundary at the same time but we conjecture that it is indeed true. The construction of the mirror coupling following the time when the two processes are simultaneously on the boundary has not been properly addressed in [BK00] and [BB99]. In an earlier paper [Wan94], mirror couplings were used without any proof of their existence. The full proof of the existence of mirror couplings in piecewise smooth domains has been given in [AB04], and the motion of the mirror following the time when both particles are on the boundary has been analyzed in [Bur05].

Next we present a general outline of a "scaling coupling" introduced by Pascu [Pas02]. See Chap. 7 for details. The main objective of any coupling technique is to construct two processes whose relative motion is highly restricted, although each of the processes by itself is a reflected Brownian motion. This can lead to a condition such as (3.6) that can be, in turn, translated into an analytic statement using a formula such as (3.4).

Pascu's idea was to start with a planar Brownian motion X_t and let $Y_t = X_{at}/\sqrt{a}$, for some fixed $a > 0$. It is well known that Y is also a planar Brownian motion. The novelty of this coupling lies in the fact that although the shape of the trajectory of Y is a scaled image of the shape of the trajectory of X, the corresponding pieces of the trajectory are traced at different times. In other words, the two processes run with different clocks. This rules out straightforward reasoning such as that in (3.3)–(3.6) but nevertheless Pascu managed to translate the information about possible geometric positions of the two processes into an analytic statement.

Two further technical aspects of scaling couplings should be mentioned here. The hot spots problem needs a construction of a pair of reflected Brownian motions in a domain D, not free Brownian motions in the whole plane. Hence, the simple scaling idea has to be modified in a way somewhat reminiscent of the way the mirror coupling in the plane is modified to handle reflected Brownian motions, because if X is a reflected Brownian motion in D then $Y_t = X_{at}/\sqrt{a}$ is not. Second, Pascu combined scaling couplings with conformal mappings in order to be able to handle arbitrary convex domains with a line of symmetry (the first step was to do the construction in a semi-disc). Conformal mappings preserve reflected Brownian motions but they require a time change. It was a crucial observation of Pascu that the time change involved in his argument had the properties needed to finish the argument when the domain was convex.

Chapter 4
Neumann Eigenfunctions and Eigenvalues

This chapter is based on [BB99]. We will discuss some properties of Neumann eigenfunctions needed in the context of the hot spots problem. Let $p_t(x, y)$ denote the Neumann heat kernel for the domain D. Under some smoothness assumptions on the domain (convex or Lipschitz boundary is sufficient by [BH91] or [Dav89, Theorems 1.7.9 and 3.2.9]), we have the following bound for the heat kernel,

$$p_t(x, y) \leq \frac{c_1}{t^{n/2}} \exp\left(-\frac{|x-y|^2}{c_2 t}\right),$$

for all $x, y \in D$ and all $0 < t \leq 1$. In particular

$$0 \leq p_1(x, y) \leq c_1. \tag{4.1}$$

Here c_1 and c_2 are constants depending on the domain. It follows from this that there are positive constants c_3 and c_4 such that

$$\sup_{x,y \in D} \left| p_t(x, y) - \frac{1}{\text{vol}(D)} \right| \leq c_3 e^{-c_4 t},$$

for $t \geq 1$ (see [BH91] or [Dav89, p. 112]). This estimate implies that $\int_D p_t(x, x)dx < \infty$. That is, the semigroup is of finite trace. By [Dav89, Theorems 1.7.9 and 1.7.12], it also has a discrete spectrum on $L^2(D)$. Let $\varphi_1, \varphi_2, \ldots$ be an orthonormal basis of $L^2(D)$ of eigenfunctions with eigenvalues $0 = \mu_1 < \mu_2 \leq \mu_3 \ldots$ where we repeat the eigenvalues if needed to take into account their multiplicity. Note that $\varphi_1 = 1/\text{vol}(D)$.

Proposition 4.1. *Let D be a domain in $\mathbb{R}^d$ whose Neumann heat kernel satisfies (4.1). Let $u_0 \in L^\infty(D)$ and let $u(t, x)$ be the solution of (3.1). Suppose that $\mu_2 = \mu_3 = \cdots = \mu_{k-1} < \mu_k$. Then,*

K. Burdzy, *Brownian Motion and its Applications to Mathematical Analysis*,
Lecture Notes in Mathematics 2106, DOI 10.1007/978-3-319-04394-4_4,
© Springer International Publishing Switzerland 2014

$$u(t, x) = \sum_{j=1}^{k-1} a_j e^{-\mu_2 t} \varphi_j(x) + R(t, x),$$

and there is a constant C depending on D, u_0 and k such that

$$|R(t, x)| \le C e^{-\mu_k t},$$

for all $t \ge 2$ and all $x \in D$.

See [BB99] for the proof.

Lemma 4.2. *Suppose that a domain D in $\mathbb{R}^2$ is divided by a smooth curve Γ into two subdomains D_1 and D_2. Let λ_j be the first eigenvalue for the mixed Neumann-Dirichlet problem on D_j, with the Neumann boundary conditions on $\partial D \cap \partial D_j$ and the Dirichlet boundary conditions on Γ. If μ_2 is the second Neumann eigenvalue for D then $\mu_2 \le \max\{\lambda_1, \lambda_2\}$.*

See [BB99] for the proof.

The nodal set of an eigenfunction u is the set of all points in $\overline{D}$ where the function u vanishes. By the Courant Nodal Line Theorem ([Cha84, p. 19] or [Ban80, p. 112]), the nodal set of a second eigenfunction for a domain in $\mathbb{R}^2$ is a smooth curve, called the nodal line, dividing the domain into two subdomains. In the case of a Neumann second eigenfunction, there are no closed nodal lines [Ban80, p. 128]. Note that if Γ is the nodal line for a second Neumann eigenfunction in D, then $\mu_2 = \lambda_1 = \lambda_2$, in the notation of Lemma 4.2.

Recall that $\mathcal{B}(x, r)$ denotes the open ball with center x and radius r. The first Dirichlet eigenvalue for $\mathcal{B}(x, r)$ is $j_0^2 / 2r^2$ where j_0 is the smallest positive zero of the first Bessel function [Ban80, p. 92].

Proposition 4.3. *Suppose that D is a planar domain with piecewise smooth boundary, $z_L, z_R \in \overline{D}$, $\rho > 0$, and a smooth curve Γ divides D into two subdomains D_1 and D_2 with $z_L \in \overline{D}_1$ and $z_R \in \overline{D}_2$. Assume that the distance from z_L to Γ is greater than or equal to ρ, and the same for z_R. Suppose that $\mathcal{B}(z_L, \rho) \cap D_1$ and $\mathcal{B}(z_R, \rho) \cap D_2$ are star-shaped domains with respect to z_L and z_R, respectively. If μ_2 is the second Neumann eigenvalue for D then $\mu_2 \le j_0^2 / 2\rho^2$.*

Proof. By Lemma 4.2, it is enough to prove that $\max\{\lambda_1, \lambda_2\} \le \lambda = j_0^2 / 2\rho^2$ where λ_i is the first eigenvalue for the D_i with Dirichlet boundary conditions on Γ and Neumann conditions elsewhere on the boundary. Let $D_3 = \mathcal{B}(z_L, \rho) \cap D_1$ and let η_1 be the first eigenvalue for the domain D_3 with Dirichlet boundary conditions on $\partial_1 D_3 = \partial D_3 \cap \partial \mathcal{B}(z_L, \rho)$ and Neumann on $\partial_2 D_3 = \partial D_3 \setminus \partial_1 D_3$. By domain monotonicity, $\lambda_1 \le \eta_1$. We now prove that $\eta_1 \le \lambda$. Towards this end, let X_t be a Brownian motion in D_3 starting from a point $y \in D_3$, killed on $\partial_1 D_3$ and reflected on $\partial_2 D_3$. Without loss of generality assume that z_L is the origin. The radial component of the inward normal vector at any point of $\partial_2 D_3$ points towards the origin (or vanishes) because D_3 is star-shaped with respect to z_L. It follows that $|X_t|$

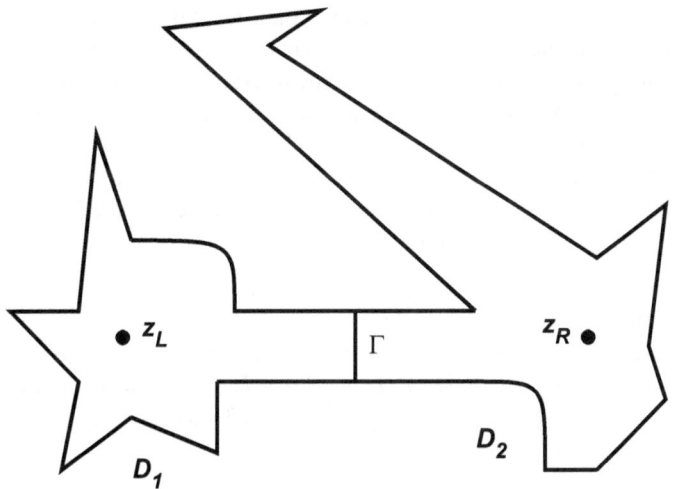

Fig. 4.1 Domain which is "locally star-shaped" with respect to two points

is a two-dimensional Bessel process plus a non-increasing process corresponding to the local time push on $\partial_2 D_3$. Hence, the time τ when the process $|X_t|$ reaches the level ρ and gets killed is not smaller than the analogous time for the two-dimensional Bessel process. The probability that the two-dimensional Bessel process does not hit ρ by the time t is the same as the probability that the exit time for disc of radius ρ is larger than t. Such probability, starting from y, is bounded below by $c(y)e^{-\lambda t}$, for large time. Here we may take $c(y)$ to be the first Dirichlet eigenfunction for the disc by verifying that the semigroup of the Dirichlet Laplacian is "intrinsically ultracontractive" and applying [Dav89, Theorem 4.2.5]. The same estimate applies to τ and it follows that $\eta_1 \leq \lambda$. A similar bound holds for D_2 and the proposition follows.   □

An example of a domain D, points z_L and z_R, and a curve Γ satisfying the assumptions of Proposition 4.3 is given in Fig. 4.1.

For any domain D in $\mathbb{R}^d$ let d_D denote its diameter.

Corollary 4.4. *Let D be a convex domain in the plane and let μ_2 be its second Neumann eigenvalue. Then $\mu_2 \leq 2j_0^2/d_D^2$.*

Proof. Consider any points $z_L, z_R \in \partial D$ with $|z_L - z_R| = d_D$. Let Γ be the intersection of D with the line of symmetry for these points. Then we can apply Proposition 4.3 with $\rho = d_D/2$.   □

The bound in Corollary 4.4 is the best possible estimate in the class of all convex planar domains. It is nearly sharp for isosceles triangles with vertices $(-1,0)$, $(1,0)$, and $(0,a)$, with very small $a > 0$. For small a, the second eigenvalue μ_2 for this triangle is bounded below by $\tan^{-1}(a) j_0^2/(2a)$, by a simple reflection argument; see [Ban80, p. 114].

Let us define the width of the domain, which we will denote by width(D), to be the infimum of the widths of all the strips containing D. We let w_D^V be the length of the projection of D on the vertical axis and w_D^H be the length of its projection on the horizontal axis. We assume, without loss of generality, that our domains are always oriented so that $w_D^V \leq w_D^H$ and we set $w_D = w_D^V$. Observe that $w_D \geq$ width(D) but in general w_D is not necessarily the same as width(D). We will say that a function $h(x^1, x^2)$ is antisymmetric with respect to the horizontal axis if $h(x^1, x^2) = -h(x^1, -x^2)$ for all (x^1, x^2).

Proposition 4.5. *(i) Suppose that D is a convex domain with the ratio $d_D/$ width(D) greater than $4j_0/\pi \approx 3.06$. Then there exists only one eigenfunction corresponding to μ_2, up to a multiplicative constant.*

(ii) Suppose that D is a convex domain which is symmetric with respect to the horizontal axis. If $d_D/w_D > 2j_0/\pi \approx 1.53$ then the subspace of $L^2(D)$ corresponding to μ_2 is one-dimensional.

Proof. (i) Assume without loss of generality that D is oriented in such a way that its projection on the vertical axis is equal to width(D). Choose points $z_L = (z_L^1, z_L^2)$ and $z_R = (z_R^1, z_R^2)$ in ∂D with the smallest z_L^1 and the largest z_R^1. The choice might not be unique and it is not necessarily true that $|z_L - z_R| = d_D$ or that z_L and z_R lie on the horizontal axis. Suppose that there exist two independent eigenfunctions $\hat{\varphi}_2(x)$ and $\tilde{\varphi}_2(x)$ corresponding to the second eigenvalue μ_2. First we will show that there exists an eigenfunction $\varphi_2(x)$ corresponding to μ_2 and such that $\varphi_2(z_L) = 0$. If $\hat{\varphi}_2(z_L) = 0$ or $\tilde{\varphi}_2(z_L) = 0$ then we are done. Otherwise we let

$$\varphi_2(x) = \hat{\varphi}_2(z_L)\tilde{\varphi}_2(x) - \tilde{\varphi}_2(z_L)\hat{\varphi}_2(x).$$

Recall that the nodal line Γ for $\varphi_2(x)$ divides D into two subdomains D_1 and D_2, and does not form a closed loop. One of the endpoints of Γ is z_L; let the other be called v. Without loss of generality we will assume that v lies on the lower part of the boundary, between z_L and z_R (we may have $v = z_R$). Let D_1 be the subdomain which lies "below" Γ. The function $\varphi_2(x)$, restricted to D_1, is the first eigenfunction for the mixed Neumann-Dirichlet problem in D_1, with the Neumann boundary conditions on $\Lambda = \partial D \cap \partial D_1$ and Dirichlet boundary conditions on Γ. We will estimate the first eigenvalue for this problem, which is the same as μ_2.

Let $a_1 = \inf\{y^2 : (y^1, y^2) \in D\}$ and $a_2 = \sup\{y^2 : (y^1, y^2) \in D\}$. Let $X_t = (X_t^1, X_t^2)$ be a Brownian motion in D_1, starting from $(b, a) \in D$, which is reflected on Λ and killed upon hitting Γ. Note that the vertical component of the inward normal vector points upward at every point of Λ. Thus, the process X_t^2 is the sum of a one-dimensional Brownian motion and a non-decreasing process, corresponding to the upward component of reflection when X_t is reflecting on Λ. The process X_t^2 cannot take values outside $[a_1, a_2]$, it is killed at the hitting time of a_2 or before hitting this value, it is pushed upward when it hits a_1 and, possibly, when it is strictly inside (a_1, a_2). A standard comparison argument for the solutions of stochastic differential equations shows now that the distribution of X^2 at time t is minorized by the distribution of the one-dimensional Brownian motion in $[a_1, a_2]$, starting from a, reflecting on a_1, and killed upon hitting a_2. The probability that such a process does not exit $[a_1, a_2]$ by the time t is bounded by $ce^{-\lambda t}$, where λ is the eigenvalue for the Laplacian on $[a_1, a_2]$, with the Neumann condition at a_1 and the Dirichlet condition at the other endpoint. Hence,

$$\lambda = (\pi^2/8)/(a_2 - a_1)^2 = (\pi^2/8)/(\text{width}(D))^2.$$

For large t, the probability that X_t^2 has not hit Γ by the time t is also bounded by $ce^{-\lambda t}$, and so the first eigenvalue for the mixed problem in D_1 cannot be smaller than $(\pi^2/8)/(\text{width}(D))^2$. Recall that this eigenvalue is the same as μ_2.

By Corollary 4.4, $\mu_2 \leq 2j_0^2/d_D^2$, so we have

$$(\pi^2/8)/(\text{width}(D))^2 \leq 2j_0^2/d_D^2,$$

which gives $d_D/\text{width}(D) \leq 4j_0/\pi$. If this inequality is not satisfied, we have only one eigenfunction corresponding to μ_2.

(ii) In this part we let z_L and z_R be the points of intersection of ∂D with the horizontal axis. We assume, as in part (i), that there are two independent eigenfunctions corresponding to μ_2 and we construct an eigenfunction $\varphi_2^*(x)$ which vanishes at z_L. This means that the nodal line of $\varphi_2^*(x)$ has one of its endpoints at z_L. It follows that $\varphi_2^*(x)$ cannot be symmetric with respect to the horizontal axis. Hence, the function

$$\varphi_2(x^1, x^2) = \varphi_2^*(x^1, x^2) - \varphi_2^*(x^1, -x^2)$$

cannot be identically equal to zero. Note that $\varphi_2(x)$ is an antisymmetric eigenfunction, i.e., $\varphi_2(x^1, x^2) = -\varphi_2(x^1, -x^2)$ for all $(x^1, x^2) \in D$. Now we can repeat the proof in part (i) with $[a_1, a_2]$ replaced by $[a_1, 0]$, since we know in the present case that the nodal line for $\varphi_2(x)$ lies on the horizontal axis.

$\square$

There exist purely analytic proofs of the above results; see [BB99] for an outline.

For an arbitrary convex domain D in the plane, the number of linearly independent eigenfunctions corresponding to the second eigenvalue is at most two in the case of the Dirichlet problem [Lin87] and also in the case of the Neumann problem [Nad86, Nad87].

Problem 4.6. Does Proposition 4.5 hold under the weaker assumption that $d_D/\text{width}(D) > \sqrt{2}$?

Note that $d_D/\text{width}(D) = \sqrt{2}$ for any square and the subspace corresponding to the second Neumann eigenvalue in a square is two-dimensional.

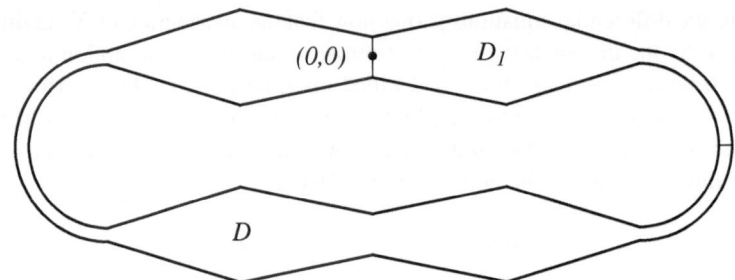

Fig. 4.2 Drawing not to scale

4.1 Eigenfunctions in Domains with Bottlenecks

This section is based on [Bur05]. We will analyze Neumann eigenfunctions and eigenvalues in a domain with a bottleneck. The methods that we will present can be applied in much more general situations but we will consider a specific domain because we will later use it to illustrate a different point.

The angle between $x = r_x e^{i\theta_x}$ and $y = r_y e^{i\theta_y}$, i.e., $\theta_x - \theta_y$, will be denoted $\angle(x, y)$. We will write $\angle(x)$ instead of $\angle(x, (1, 0))$, i.e., $\angle(x)$ will denote the angle formed by the vector x with the positive horizontal semi-axis. We will use the convention that $\angle(x, y) \in (-\pi, \pi]$.

We will now consider a domain $D \subset \mathbb{R}^2$ whose definition will involve a parameter $\varepsilon \in (0, 1/4)$. The value of ε will be chosen later and should be thought of as a very small number; it will be suppressed in the notation. Let A_1 be a convex polygonal domain with the consecutive vertices $(0, -\varepsilon)$, $(0, \varepsilon)$, $(1, 2\varepsilon)$, $(2, \varepsilon_0)$, $(2, -\varepsilon_0)$ and $(1, -2\varepsilon)$, where $\varepsilon_0 \in (0, \varepsilon)$. The value of the parameter ε_0 will be specified later—it will be much smaller than ε. Let C_1 be the (shorter) arc of $\mathcal{B}((2, -1), 1 + \varepsilon_0)$ with endpoints $(2, \varepsilon_0)$ and $(3 + \varepsilon_0, -1)$ and, similarly, let C_2 be the arc of $\mathcal{B}((2, -1), 1 - \varepsilon_0)$ with endpoints $(2, -\varepsilon_0)$ and $(3 - \varepsilon_0, -1)$. Let A_2 be an open domain whose boundary consists of C_1, C_2, and line segments $\overline{(2, \varepsilon_0), (2, -\varepsilon_0)}$ and $\overline{(3 + \varepsilon_0, -1), (3 - \varepsilon_0, -1)}$. Let $A_3 = A_1 \cup A_2$, let A_4 be the symmetric image of A_3 with respect to the line $\{(x_1, x_2) : x_2 = -1\}$, and let A_5 and A_6 be the symmetric images of A_3 and A_4 with respect to $\{(x_1, x_2) : x_1 = 0\}$. Finally we let D be the interior of the closure of $A_3 \cup A_4 \cup A_5 \cup A_6$. A schematic drawing of D is presented in Fig. 4.2.

Recall synchronous couplings defined in (3.5). Note that on any interval (s, t) such that $X_u \in D$ and $Y_u \in D$ for all $u \in (s, t)$, we have $X_u - Y_u = X_s - Y_s$ for all $u \in (s, t)$.

Recall that μ_2 denotes the second Neumann eigenfunction in D.

Lemma 4.7. *For any* $c_1 > 0$ *there exists* $c_2 > 0$ *such that if* $\varepsilon_0 \le c_2 \varepsilon$ *then* $\mu_2 \le c_1$ *and* μ_2 *is simple.*

Proof. Step 1.   Let $r = \varepsilon_0/(2\varepsilon - \varepsilon_0)$ and note that the point $y := (2+r, 0)$ lies at the intersection of straight lines passing through the line segments $\overline{(1, 2\varepsilon), (2, \varepsilon_0)}$ and $\overline{(1, -2\varepsilon), (2, -\varepsilon_0)}$. Let $K = \{x = (x_1, x_2) \in D : |x_1| \leq 2, x_2 > -1\}$, $K_1 = \mathcal{B}(y, 2r) \cap K$, $K_2 = \partial\mathcal{B}(y, 1/2) \cap K$ and $K_3 = \partial\mathcal{B}(y, 1) \cap K$. Let X be a reflected Brownian motion in D with $X_0 \in K_2$. Let $T_0 = 0$, and for $k \geq 1$ let

$$S_k = \inf\{t \geq T_{k-1} : X_t \in K_1 \cup K_3\},$$
$$T_k = \inf\{t \geq S_k : X_t \in K_2\}.$$

Let $R_t = \text{dist}(X_t, y)$ and note that if X is between K_1 and K_3, the process R is a two-dimensional Bessel process because the normal reflection of X on ∂D has no effect on R. It follows that for any $p_1 < 1$, there exists $r_0 > 0$ so small that if $r \leq r_0$ then

$$\mathbb{P}(X_{S_k} \in K_3 \mid \mathcal{F}_{T_{k-1}}) = \frac{\log(1/2) - \log(2r)}{\log 1 - \log(2r)} \geq p_1.$$

Moreover, for some $t_0 > 0$ not depending on r,

$$\mathbb{P}(X_{S_k} \in K_3, S_k - T_{k-1} > t_0 \mid \mathcal{F}_{T_{k-1}}) \geq p_1.$$

Let $z = (-2 - r, 0)$,

$$K_1' = (\mathcal{B}(y, 2r) \cup \mathcal{B}(z, 2r)) \cap K,$$
$$K_2' = (\partial\mathcal{B}(y, 1/2) \cup \partial\mathcal{B}(z, 1/2)) \cap K,$$
$$K_3' = (\partial\mathcal{B}(y, 1) \cup \partial\mathcal{B}(z, 1)) \cap K,$$

$T_0' = 0$, and for $k \geq 1$ let

$$S_k' = \inf\{t \geq T_{k-1}' : X_t \in K_1' \cup K_3'\},$$
$$T_k' = \inf\{t \geq S_k' : X_t \in K_2'\}.$$

By symmetry, $\mathbb{P}(X_{S_k'} \in K_3', S_k' - T_{k-1}' > t_0 \mid \mathcal{F}_{T_{k-1}'}) \geq p_1$. By the repeated use of the strong Markov property,

$$\mathbb{P}(T_{K_1'}^X \geq k t_0 \mid X_0 \in K_2') \geq \mathbb{P}\left(\bigcap_{1 \leq j \leq k} \{X_{S_j'} \in K_3', S_j' - T_{j-1}' > t_0\} \mid X_0 \in K_2'\right)$$
$$\geq p_1^k.$$

Let $D_- = \{x = (x_1, x_2) \in D : x_2 < -1\}$ and $A = \partial D_- \cap D$. Let $u(t, x)$ be the heat equation solution in D with the Neumann boundary conditions and the initial condition $u(0, x) = 1$ for $x \in D_-$ and $u(0, x) = 0$ otherwise. Note that u can be represented probabilistically as $u(t, x) = \mathbb{P}(X_t \in D_- \mid X_0 = x)$. By the strong Markov property applied at T_A^X and symmetry, $\mathbb{P}(X_t \in D_- \mid T_A^X < t) = 1/2$, so for $x \in K_2'$ and large t,

$$
\begin{aligned}
u(t, x) &= \mathbb{P}(X_t \in D_- \mid X_0 = x) \\
&= \mathbb{P}(X_t \in D_-, T_A^X < t \mid X_0 = x) \\
&= (1/2)\, \mathbb{P}(T_A^X < t \mid X_0 = x) \\
&= 1/2 - (1/2)\, \mathbb{P}(T_A^X \geq t \mid X_0 = x) \\
&\leq 1/2 - (1/2)\, \mathbb{P}(T_{K_1'}^X \geq t \mid X_0 = x) \\
&\leq 1/2 - (1/2) p_1^{t/(2t_0)} = 1/2 - (1/2) e^{(\log p_1/(2t_0))t}.
\end{aligned}
$$

Since p_1 can be made arbitrarily close to 1 by making r small, $-\log p_1/(2t_0) > 0$ can be arbitrarily close to 0. By symmetry, $u(t, x)$ converges to $1/2$ as $t \to \infty$. By Proposition 4.1, $\sup_{x \in D} |u(t, x) - 1/2| \leq c_3 e^{-\mu_2 t}$ for large t. Hence, $\mu_2 \leq -\log p_1/(2t_0)$ and we see that for any $c_1 > 0$ we have $\mu_2 \leq c_1$, provided r is sufficiently small. If $c_2 < 1$ and $\varepsilon_0 \leq c_2 \varepsilon$ then $r = \varepsilon_0/(2\varepsilon - \varepsilon_0) \leq c_2$, so $\mu_2 \leq c_1$ if we assume that c_2 is small. This proves the first claim of the lemma.

Step 2. The rest of the proof is based on [BW99]. In this step we will show that the nodal lines of any eigenfunction corresponding to μ_2 are confined to a small subset of D when ε_0 is small.

Consider any second Neumann eigenfunction φ_2 in D. Let Γ be its nodal line, i.e., $\Gamma = \{x \in D, \varphi_2(x) = 0\}$ (note that the line Γ is not necessarily connected). Let $M = \{(x_1, x_2) \in D : |x_1| > 3/2\}$. Our next goal is to prove that for all small enough $\varepsilon > 0$, $\Gamma \subset M$. Let Γ_1 denote a connected component of the nodal line. Suppose that Γ_1 intersects $D \setminus M$ and that the diameter of Γ_1 is less than $\varepsilon/2$. We will show that this assumption leads to a contradiction, if ε_0 is sufficiently small. As the diameter of Γ_1 is less than $\varepsilon/2$ and $\Gamma_1 \not\subset M$, it is easy to see that Γ_1 has to cut off a domain D_4 from D of diameter less than ε (the boundary of D_4 would consist of Γ_1 and a piece of ∂D). It is also easy to prove that the first eigenvalue λ_1 for the mixed problem in D_4, with the Dirichlet conditions on Γ_1 and the Neumann conditions elsewhere on ∂D_4, is larger than some $\lambda_0 > 0$, independent of ε and the shape and location of Γ_1 (but using the fact that Γ_1 intersects $D \setminus M$). Since $\lambda_1 = \mu_2$, we can adjust ε_0 to make $\mu_2 < \lambda_0$ using Step 1, and we can thus rule out the possibility that the diameter of Γ_1 is less than $\varepsilon/2$.

Step 3. It is easy to see that there exists $p_1 > 0$ such that for all ε, for all $x \in M$,

$$
\mathbb{P}(T(D \setminus M) < 1/2 \mid X_0 = x) > p_1.
$$

Homework 4.8. Suppose that $\gamma \subset D$ is a connected set of diameter greater than $\varepsilon/2$ such that $\gamma \not\subset M$. Show that there exists $p_2 > 0$ such that for all small ε and ε_0, for all $x \in D \setminus M$,

$$\mathbb{P}(T_\gamma < 1/2 \mid X_0 = x) \geq p_2.$$

Step 4.   Assume that $\Gamma \not\subset M$. By the Courant Nodal Line Theorem, the nodal line Γ divides D into two connected components. Let Γ_1 denote a connected component of Γ that intersects $D \setminus M$. Step 2 implies that the diameter of Γ_1 is at least $\varepsilon/2$. Hence, using the results of Step 3, we get that for all ε and ε_0, for all $x \in D \setminus M$,

$$\mathbb{P}(T_\Gamma \leq 1/2 \mid X_0 = x) \geq p_2.$$

On the other hand, for all $x \in M$, using the strong Markov property at time $T_{D\setminus M}$, Step 3 and the last inequality, we get that

$$\mathbb{P}(T_\Gamma \leq 1 \mid X_0 = x) \geq p_2 p_1.$$

Hence, for all $x \in D$,

$$\mathbb{P}(T_\Gamma \leq 1 \mid X_0 = x) \geq p_1 p_2,$$

so that the Markov property applied at times $n = 1, 2, \ldots$, implies that for all $n \geq 1$,

$$\mathbb{P}(T_\Gamma \geq n \mid X_0 = x) \leq (1 - p_1 p_2)^n$$

and consequently that $\mu_2 \geq -\log(1 - p_1 p_2)$. Note that p_1 and p_2 are independent of ε. Hence, combining this with Step 1 shows that for small enough ε and ε_0, one never has $\Gamma \not\subset M$. In the rest of the proof $\varepsilon, \varepsilon_0 > 0$ are assumed to be small enough so that the nodal line of any second Neumann eigenfunction in D is a subset of M.

Step 5.   We will first prove the following very general claim. If there exists $z_0 \in D$ such that the nodal line of any second Neumann eigenfunction does not contain z_0 then the second eigenvalue is simple.

To prove the claim, suppose that φ_2 and $\tilde{\varphi}_2$ are two independent eigenfunctions corresponding to μ_2. By assumption, $\varphi_2(z_0) \neq 0$ and $\tilde{\varphi}_2(z_0) \neq 0$ so the function

$$x \mapsto \varphi_2(x)\tilde{\varphi}_2(z_0) - \varphi_2(z_0)\tilde{\varphi}_2(x)$$

is a non-zero eigenfunction corresponding to μ_2. Since it vanishes at z_0, we obtain a contradiction.

We can apply the claim to our domain because $\Gamma \subset M$. We conclude that μ_2 is simple.

$\square$

Chapter 5
Synchronous and Mirror Couplings

This chapter is based on [AB04]. Recall from Sect. 3.1 that a planar set D is called a *lip domain* if it is Lipschitz, open, bounded, connected, and given by

$$D = \{(x_1, x_2) : f_1(x_1) < x_2 < f_2(x_1)\}, \tag{5.1}$$

where f_1, f_2 are Lipschitz functions with constant 1.

Homework 5.1. The assumption that D is a Lipschitz domain puts an extra constraint on the functions f_k besides that they are Lipschitz. Discuss the extra constraint.

Reflected Brownian motion can be constructed in an arbitrary open set D; more precisely, it can be constructed on the Martin-Kuramochi compactification of D. The first construction was given by Fukushima in [Fuk67]. Heuristically, one can think about reflected Brownian motion in an arbitrary open set as the limit, in distribution, of reflected Brownian motions in approximating smooth domains (see [BC98]).

Suppose that D is a lip domain and let B denote two-dimensional Brownian motion. Since the unit inward normal vector $\mathbf{n}(z)$ is well defined for almost all $z \in \partial D$, it makes sense to consider the following representation of reflected Brownian motion

$$X_t = x + B_t + \int_0^t \mathbf{n}(X_s) dL_s^X,$$

where $x \in \overline{D}$ and L_s^X is the local time of X on the boundary of D, i.e., L^X is a non-decreasing process that does not increase when X is inside D. In other words, $\int_0^\infty \mathbf{1}_D(X_s) dL_s^X = 0$. This SDE has a unique pathwise solution in lip domains by the results in [BBC05]. It follows that the following pair of SDE's has a unique strong solution, for $x, y \in \overline{D}$,

K. Burdzy, *Brownian Motion and its Applications to Mathematical Analysis*, Lecture Notes in Mathematics 2106, DOI 10.1007/978-3-319-04394-4_5, © Springer International Publishing Switzerland 2014

$$X_t = x + B_t + \int_0^t \mathbf{n}(X_s)dL_s^X, \tag{5.2}$$

$$Y_t = y + B_t + \int_0^t \mathbf{n}(Y_s)dL_s^Y.$$

We call (X, Y) a *synchronous coupling* of reflected Brownian motions.

"Mirror couplings" of reflected diffusions were applied in [Wan94, BB99] and [BK00]. Unfortunately, none of these papers paid sufficient attention to the construction of this type of couplings. The paper [Wan94] does not offer any proof of existence of mirror couplings for reflected diffusions. The papers [BB99] and [BK00] give an explicit construction for the mirror coupling of Brownian motions reflected on a straight line. This construction breaks down in polygonal domains when two Brownian particles hit two different line segments on the boundary at the same time.

Homework 5.2 (Hard). Prove that, in some polygonal domains, reflected Brownian motions which form a mirror coupling can hit the boundary of the domain simultaneously at the same instant before the coupling time, with positive probability.

5.1   Construction of Mirror Couplings

This section is a cautionary tale. It shows that it takes a considerable effort to prove existence and determine properties of mirror couplings even in reasonably smooth domains.

Suppose for a moment that $D \in \mathbb{R}^d$, $d \geq 2$, is a smooth domain. We are looking for a pair of reflected Brownian motions (X_t, Y_t) satisfying the following system of stochastic differential equations,

$$dX = dW + dL, \quad dY = dZ + dM, \quad dZ = dW - 2m\, m \cdot dW, \quad m = \frac{Y - X}{|Y - X|},$$

for times less than $\zeta = \inf\{s : X_s = Y_s\}$. Here W is a d-dimensional Brownian motion, and Z is another Brownian motion for which the increments are mirror images of those of W, the mirror being the $(d - 1)$-dimensional hyperplane with respect to which X and Y are symmetric. The processes X and Y are reflected Brownian motions with boundary terms L and M respectively. We will prove pathwise uniqueness and strong existence for this system of equations. As a corollary, it will be proved that (X_t, Y_t) is strong Markov. The final proposition in this section will show that mirror couplings in lip domains have an "order preserving" property.

5.2 An SDE with Reflecting Boundary Conditions

The starting point for our analysis are some results from [LS84] regarding unique solvability of an SDE with reflection. The results in that paper apply to a class of domains satisfying complicated assumptions—we refer the reader to [LS84] for details. It will suffice to say that the results of [LS84] apply to all $D \subset \mathbb{R}^d$, $d \geq 3$, which are C^2-smooth and all planar piecewise C^2-smooth domains with a finite number of convex corners (more precisely, domains for which the boundary consists of finitely many smooth parts, the uniform exterior sphere condition and the uniform interior cone condition are satisfied; see [LS84]).

Fix a bounded open set $D \in \mathbb{R}^d$ for some $d \geq 2$ and assume that its boundary is C^2-smooth if $d \geq 3$ or piecewise smooth with a finite number of convex corners, if $d = 2$. For $x \in \partial D$, let $\mathbf{n}(x)$ be the inward unit normal vector, if it exists. If x is a corner, we let $\mathbf{n}(x)$ to be an arbitrary unit vector pointing inside the domain at x (this is just for the sake of completeness of the definition—the reflected Brownian motion does not visit corners with probability 1 so the definition of $\mathbf{n}(x)$ for such points is irrelevant).

Let $(\Omega, \mathcal{F}, (\mathcal{F}_t), \mathbb{P})$ be a complete filtered probability space, on which an $(\mathcal{F}_t)$-Brownian motion (W_t) is given. It was shown in [LS84] that for any $x \in \overline{D}$ there exists a unique continuous $(\mathcal{F}_t)$-semimartingale (X_t) with values in $\overline{D}$ for all $t \geq 0$, a.s., satisfying

$$X_t = x + W_t + L_t, \quad |L|_t = \int_0^t 1_{X_s \in \partial D} d|L|_s, \quad L_t = \int_0^t \mathbf{n}(X_s) d|L|_s, \quad (5.3)$$

where L_t is a continuous process with values in $\mathbb{R}^d$, with variation $|L|_t$ which is bounded on each finite interval. In fact, there is a map $\Gamma : C([0, \infty) : \mathbb{R}^d) \to C([0, \infty) : \mathbb{R}^d)$ (the "Skorohod map") such that $X = \Gamma(x + W)$ a.s. Moreover, for every $T > 0$, there exists a map Γ_T from $C([0, T] : \mathbb{R}^d)$ into itself, such that

$$X|_{[0,T]} = \Gamma(x + W)|_{[0,T]} = \Gamma_T(x + W|_{[0,T]}), \quad (5.4)$$

and Γ_T is Hölder continuous of order $1/2$ on compact subsets of $C([0, T] : \mathbb{R}^d)$. This implies that the function $x \mapsto \Gamma_T(x + W|_{[0,T]})$ is continuous. With an abuse of notation, we will refer below to Γ_T as Γ. We will call the process X a reflected Brownian motion in D (with normal reflection), driven by W, starting from x.

Denote

$$\|Z\|_t = [\mathbb{E} \sup_{0 \leq s \leq t} |Z_s|^4]^{1/4}.$$

Let $\mathbf{Z}$ be the space of continuous adapted processes Z for which $\|Z\|_t < \infty$ for all $t > 0$. We shall need the following version of Lemma 3.1 of [LS84].

Lemma 5.3. *Let σ be an $(\mathcal{F}_t)$-stopping time. Then for each $T > 0$ there exists C such that for all $t \leq T$ and all $Z, Z' \in \mathbf{Z}$,*

$$\|\Gamma(Z)_{\cdot \wedge \sigma} - \Gamma(Z')_{\cdot \wedge \sigma}\|_t^4 \leq C \int_0^t \|Z_{\cdot \wedge \sigma} - Z'_{\cdot \wedge \sigma}\|_s^4 ds.$$

Proof. The only difference between our lemma and Lemma 3.1 of [LS84] is that the processes are stopped at a stopping time σ in our version of the inequality. The original proof given in [LS84] applies, with minor adjustments needed to take into account the presence of the stopping time σ.   □

5.3   Equations for the Mirror Coupling

We will formulate a system of stochastic differential equations for a pair of "mirror coupled" reflected Brownian motions, i.e., satisfying the following condition. On any time interval $[s, t]$, on the event A that both processes do not hit the boundary during $[s, t]$, there is a $(d - 1)$-dimensional hyperplane (the "mirror"), depending on $\omega \in A$ and s, t, with respect to which the processes are symmetric at every time in $[s, t]$.

Let $(\Omega, \mathcal{F}, (\mathcal{F}_t), \mathbb{P}, W)$ be given, where W is a Brownian motion. We are looking for processes Z, X and Y with the following properties. The processes X and Y are reflected Brownian motions in D, starting from x and y, and driven by W and Z, respectively. Here Z is another Brownian motion on the same probability space. We want to have

$$X = \Gamma(x + W), \tag{5.5}$$

as in (5.3). We also need,

$$Y = \Gamma(y + Z). \tag{5.6}$$

Let m_t be a unit vector perpendicular to the mirror with respect to which X_t and Y_t are symmetric, namely $m_t = (Y_t - X_t)/|Y_t - X_t|$. Here and later, all vectors will be column vectors. For m in the unit sphere of $\mathbb{R}^d$, let $H(m)$ denote the $d \times d$ matrix

$$H(m) = I - 2mm', \tag{5.7}$$

where m' denotes the transpose of m. Then $H(m)v = v - 2(m \cdot v)m$ is the mirror image of v about the hyperplane through the origin, perpendicular to m. We would like Z to depend on W and on the mirror, in such a way that, before X and Y first meet, Z's increments are mirror images of W's:

$$Z_t = \int_0^t H(m_s)dW_s.$$

For $x \in \mathbb{R}^d$, let $G(x) = H(x/|x|)$ if $x \neq 0$, and let $G(0) = 0$ (the value of $G(0)$ is in fact irrelevant). Equip the space of $d \times d$ matrices with the norm $\|A\|^2 = \sum_{i,j} A_{i,j}^2$. Then it is easy to check that for $|m| = |\tilde{m}| = 1$,

$$\|H(m) - H(\tilde{m})\| \leq c|m - \tilde{m}|. \tag{5.8}$$

Consider the equation

$$Z_t = \int_0^{t \wedge \zeta} G(\Gamma(y + Z)_s - X_s)dW_s + 1_{t \geq \zeta}(W_t - W_\zeta), \quad \zeta = \inf\{s : \Gamma(y + Z)_s = X_s\}. \tag{5.9}$$

Definition 5.4. (i) We say that *pathwise uniqueness holds for Eq. (5.9)* if whenever $(\Omega, \mathcal{F}, (\mathcal{F}_t), \mathbb{P})$ is a filtered probability space, W is an $(\mathcal{F}_t)$-Brownian motion, $X = \Gamma(x + W)$, processes Z and Z' are $(\mathcal{F}_t)$-adapted and have continuous sample paths, and both Z and Z' satisfy (5.9) for $t \geq 0$ a.s., then $Z(t) = Z'(t)$ for all $t \geq 0$, a.s.

(ii) A *strong solution* to (5.9) on a given probability space $(\Omega, \mathcal{F}, \mathbb{P})$ relative to the Brownian motion W, is a process Z with continuous sample paths, adapted to the (augmented) filtration $(\mathcal{F}_t^W)$, generated by W, and with $X = \Gamma(x + W)$, which satisfies (5.9) for all $t \geq 0$, a.s.

Theorem 5.5. *Pathwise uniqueness holds for (5.9). Let W be a Brownian motion on a complete probability space $(\Omega, \mathcal{F}, \mathbb{P})$. Then there exists a strong solution of (5.9) relative to W.*

We will refer to the pair (X, Y) as a *mirror coupling* of reflected Brownian motions.

Proof. Pathwise uniqueness. We will first consider the problem on a finite time interval $[0, T]$. Let a Brownian motion W on $(\Omega, \mathcal{F}, (\mathcal{F}_t), \mathbb{P})$ be given, and let $X = \Gamma(x + W)$. Assume Z and $\tilde{Z}$ are two $(\mathcal{F}_t)$-adapted processes with continuous sample paths, satisfying (5.9) for all $t \in [0, T]$ a.s. Denote $Y = \Gamma(y + Z)$, $\tilde{Y} = \Gamma(y + \tilde{Z})$, $V = Y - X$, $\tilde{V} = \tilde{Y} - X$. Let $\tau_n = \inf\{t : |V_t| \leq 1/n\}$, $\tilde{\tau}_n = \inf\{t : |\tilde{V}_t| \leq 1/n\}$, $S_n = \tau_n \wedge \tilde{\tau}_n$. Then $S_n \uparrow \zeta \wedge \tilde{\zeta}$. To prove pathwise uniqueness it is enough to show that $\mathbb{P}(Z_t = \tilde{Z}_t, 0 \leq t < \zeta \wedge \tilde{\zeta}) = 1$. On $[0, S_n]$, one has

$$\left| \frac{V}{|V|} - \frac{\tilde{V}}{|\tilde{V}|} \right| \leq c \frac{|V - \tilde{V}|}{|V| \wedge |\tilde{V}|} \leq cn|V - \tilde{V}| = cn|Y - \tilde{Y}|. \tag{5.10}$$

We will use the following version of the Burkholder-Davis-Gundy inequality [KS91, p. 163],

$$\mathbb{E} \left| \int_0^T A_t dW_t \right|^{2m} \leq c_m T^{m-1} \mathbb{E} \int_0^T |A_t|^{2m} dt, \tag{5.11}$$

for $m \geq 1$, and A_t adapted. Since Z is adapted, it follows from (5.4) that so is the integrand in (5.9), and therefore Z is a martingale. Doob's inequality for the martingale $Z - \tilde{Z}$, along with Lemma 5.3, (5.8), (5.10) and (5.11) yield for any $t \in (0, T]$,

$$\mathbb{E} \sup_{0 \leq s \leq t} |Z_{s \wedge S_n} - \tilde{Z}_{s \wedge S_n}|^4 \leq c\, \mathbb{E}\, |Z_{t \wedge S_n} - \tilde{Z}_{t \wedge S_n}|^4$$

$$\leq cT\, \mathbb{E} \int_0^{t \wedge S_n} \|G(Y_s - X_s) - G(\tilde{Y}_s - X_s)\|^4 ds$$

$$= cT\, \mathbb{E} \int_0^{t \wedge S_n} \left\| H\left(\frac{V_s}{|V_s|}\right) - H\left(\frac{\tilde{V}_s}{|\tilde{V}_s|}\right) \right\|^4 ds$$

$$\leq cT\, \mathbb{E} \int_0^{t \wedge S_n} \left| \frac{V_s}{|V_s|} - \frac{\tilde{V}_s}{|\tilde{V}_s|} \right|^4 ds$$

$$\leq cTn^4\, \mathbb{E} \int_0^{t \wedge S_n} |Y_s - \tilde{Y}_s|^4 ds$$

$$\leq cTn^4 \int_0^t \mathbb{E} \sup_{0 \leq u \leq s} |Y_{u \wedge S_n} - \tilde{Y}_{u \wedge S_n}|^4 ds$$

$$\leq cT^2 n^4 \int_0^t \mathbb{E} \sup_{0 \leq u \leq s} |Z_{u \wedge S_n} - \tilde{Z}_{u \wedge S_n}|^4 ds.$$

Gronwall's inequality now shows that $\mathbb{E} \sup_{0 \leq s \leq T} |Z_{s \wedge S_n} - \tilde{Z}_{s \wedge S_n}|^4 = 0$. Hence $\mathbb{P}(Z_s = \tilde{Z}_s, 0 \leq s \leq S_n) = 1$, for all n, and therefore $\mathbb{P}(Z_s = \tilde{Z}_s, 0 \leq s < \zeta \wedge \tilde{\zeta}) = 1$. This shows that Z and $\tilde{Z}$ are indistinguishable on $[0, T]$. Since T is arbitrary, the same is true on $[0, \infty)$. This completes the proof of pathwise uniqueness.

Existence of Strong Solutions. By (5.8), the restriction of the matrix-valued mapping $x \mapsto G(x)$ to $\mathbb{R}^d \setminus \mathcal{B}(0, \varepsilon)$ is globally Lipschitz with respect to $\| \cdot \|$. Therefore, for every $n \in \mathbb{N}$, one can find a globally Lipschitz function G^n on $\mathbb{R}^d$ that agrees with G outside $\mathcal{B}(0, 1/n)$. Let $\lambda_n < \infty$ denote the Lipschitz constant of G^n. We are given a probability space $(\Omega, \mathcal{F}, \mathbb{P})$ with a Brownian motion W and the filtration $(\mathcal{F}_t^W)$ generated by W. Denote $X = \Gamma(x + W)$. We will first show that for any n there exists a continuous adapted process Z^n such that

$$Z_t^n = \int_0^t G^n(\Gamma(y + Z^n)_s - X_s) dW_s. \tag{5.12}$$

Pathwise uniqueness holds for (5.12). The proof is similar to that of pathwise uniqueness for (5.9).

In this paragraph we drop the dependence on n in the notation of the processes involved. Recall the space $\mathbf{Z}$ defined before the statement of Lemma 5.3. Let F map $Z \in \mathbf{Z}$ to

$$F(Z)_t = \int_0^t G^n(\Gamma(y + Z)_s - X_s) dW_s.$$

Suppose $Z, Z' \in \mathbf{Z}$. By Lemma 5.3, Doob's inequality for the martingale $F(Z) - F(Z')$ and (5.11), we have for $t \in (0, T]$,

$$\mathbb{E} \sup_{s \le t} |F(Z)_s - F(Z')_s|^4 \le c \mathbb{E} |F(Z)_t - F(Z')_t|^4$$

$$\le ct \, \mathbb{E} \int_0^t \|G^n(\Gamma(y + Z)_s - X_s)$$

$$- G^n(\Gamma(y + Z')_s - X_s)\|^4 ds$$

$$\le c\lambda_n^4 t \, \mathbb{E} \int_0^t |\Gamma(y + Z)_s - \Gamma(y + Z')_s|^4 ds$$

$$\le c\lambda_n^4 T^2 \int_0^t \mathbb{E} \sup_{0 \le u \le s} |Z_u - Z'_u|^4 ds. \tag{5.13}$$

Let $Z^{(0)} = \mathbf{0}$, and for $k \in \mathbb{N}$ let $Z^{(k)} = F(Z^{(k-1)})$. Since G^n is bounded, there is c such that $\|Z^{(1)} - Z^{(0)}\|_T \le c$. Iterating (5.13) it easily follows that

$$\|Z^{(k+1)} - Z^{(k)}\|_T^4 \le c \frac{c_1^k}{k!},$$

where c_1 is a constant that may depend on n and T. Hence by Chebyshev's inequality,

$$\mathbb{P}(\sup_{s \le T} |Z_s^{(k+1)} - Z_s^{(k)}| \ge 2^{-k}) \le c(16c_1)^k / k!,$$

and the Borel-Cantelli lemma shows that with probability one, $\sup_{0 \le s \le T} |Z_s^{(k+1)} - Z_s^{(k)}| \le 2^{-k}$ for all sufficiently large k. Hence the paths of $Z^{(k)}$ converge in the uniform topology, a.s., say, to Z. Using again (5.13), we have $\|F(Z^{(k)}) - F(Z)\|_T^4 \le c\|Z^{(k)} - Z\|_T^4$, and as a result $F(Z^{(k)})$ converge a.s. to $F(Z)$. We see that Z is a fixed point of F. Since T is arbitrary we conclude that there exists a continuous, $(\mathcal{F}_t^W)$-adapted process satisfying (5.12) for all $t \in [0, \infty)$ a.s.

We reintroduce the dependence on n in our notation. Let Z^n denote the process in (5.12) and set $Y^n = \Gamma(y + Z^n)$. Let $\tau_n = \inf\{s : |Y_s^n - X_s| \le 1/n\}$. Since we have uniqueness for (5.12), it is clear that on $[0, \tau_n]$, the processes Z^n and Z^{n+1} agree a.s. Setting $\zeta = \lim_n \tau_n$ and defining

$$Z_t = 1_{t < \zeta} \lim_{n \to \infty} Z^n(t) + 1_{t \ge \zeta}(W_t - W_\zeta),$$

we obtain a continuous and adapted process Z. We see that (5.9) is satisfied for all $t \in [0, \infty)$ a.s. Therefore there exists a strong solution relative to W. $\qquad\square$

5.4   Strong Markov Property

Let $(\Omega, \mathcal{F}, (\mathcal{F}_t), \mathbb{P}, W)$ be a complete filtered probability space and a Brownian motion. Let $U = (X, Y)$ denote the unique solution to (5.5), (5.6) and (5.9). For $u = (x, y) \in (\overline{D})^2$, let $\mathbb{P}^u$ denote the measure induced by U on $(C([0, \infty) : \mathbb{R}^{2d})$, $\mathcal{B}(C([0, \infty) : \mathbb{R}^{2d})))$, assuming $U_0 = u$. Let $\mathbb{E}^u$ denote the expectation with respect to $\mathbb{P}^u$. The transition function for the process U is defined as

$$P_t(u, f) = \mathbb{E}^u(f(U_t)),$$

for all $u \in (\overline{D})^2$ and bounded Borel measurable functions f. Given the results on existence and uniqueness, and the properties of the Skorohod map quoted before, we have the following.

Corollary 5.6. *The process U is strong Markov, i.e., for any a.s. finite $(\mathcal{F}_t)$-stopping time T and bounded Borel function f, one has $\mathbb{P}$-a.s., for all $s \geq 0$,*

$$\mathbb{E}[f(U_{T+s}) \mid \mathcal{F}_T] = P_s(U_T, f).$$

Proof. We take here the approach of [Pro90, Theorem V.32]. For each finite stopping time T, note that $\tilde{W}_s = W_{T+s} - W_T$, $s \geq 0$ is a Brownian motion (cf., e.g., [Pro90, Theorem I.32]), and consider the unique solution $\tilde{Z}$ to the following equation, with $u = (x, y) \in (\overline{D})^2$,

$$\tilde{Z}_t = \int_0^{t \wedge \tilde{\xi}} G(\Gamma(y + \tilde{Z})_s - \Gamma(x + \tilde{W})_s) d\tilde{W}_s + 1_{t \geq \tilde{\xi}}(\tilde{W}_t - \tilde{W}_{\tilde{\xi}}), \qquad (5.14)$$

where $\tilde{\xi} = \inf\{s : \Gamma(y + \tilde{Z})_s = \Gamma(x + \tilde{W})_s\}$. To take into account the dependence on the stopping time and the initial condition, we denote $\tilde{Z}_t$ above by $Z(u, T, t)$, and we also let

$$U(u, T, t) = (\Gamma(x + W_{\cdot + T})_t, \Gamma(y + Z(u, T, \cdot))_t).$$

Denote the Borel sigma-field of subsets of $(\overline{D})^2$ [resp., $\mathbb{R}_+$] by $\mathcal{U}$ [resp., $\mathcal{R}_+$]. We will show that, for a given stopping time, the mapping $(u, t, \omega) \mapsto Z(u, T, t)$ has a $\mathcal{U} \otimes \mathcal{R}_+ \otimes \mathcal{F}$ jointly measurable version, and hence so does $U(u, T, t)$. We will use these versions in the rest of the proof. Recall the processes $Z^{(k)}$ and Z^n of the proof of Theorem 5.5. Recall also that for each n, $Z^{(k)}$ converge uniformly on compacts in probability to Z^n. Since each $Z^{(k)}$ is jointly measurable, this implies that so is Z^n (cf. [Pro90]). Since Z is the pointwise limit of Z^n as $n \to \infty$, this shows that Z is jointly measurable. The same argument applies to $Z(u, T, t)$ for any fixed stopping time T. A similar measurability property holds for $U(u, T, t)$ by the uniform continuity of Γ on compact sets.

Fix a finite stopping time T and set $\mathcal{F}^* = \sigma\{W_{T+u} - W_T : u \geq 0\}$. Then $\mathcal{F}^*$ is independent of $\mathcal{F}_T$ under $\mathbb{P}$, since W is an $(\mathcal{F}_t)$-Brownian motion. Note that $Z(u, T, t)$ is measurable with respect to $\mathcal{F}^*$. By the uniqueness results for the Skorohod map (the uniqueness holds in the deterministic sense), $\Gamma(R)_{T+s} = \Gamma(\Gamma(R)_T - R_T + R_{T+.})_s$, $s \geq 0$, whenever R is an $(\mathcal{F}_t)$-Brownian motion. Fix a u for the moment and denote $Z_t = Z(u, 0, t)$. Then we have for $t \geq 0$, on the event $\{\zeta > T + t\}$,

$$
Z_{T+t} - Z_T = \int_T^{T+t} G(\Gamma(y + Z)_s - \Gamma(x + W)_s) dW_s
$$
$$
= \int_0^t G(\Gamma(\Gamma(y + Z)_T + (Z_{T+.} - Z_T))_s
$$
$$
- \Gamma(\Gamma(x + W)_T + \tilde{W})_s) d\tilde{W}_s. \tag{5.15}
$$

Let $\bar{\zeta} = \inf\{s \geq 0 : \Gamma(y + Z_{T+.})_s = \Gamma(x + W_{T+.})_s\}$, and note that $\bar{\zeta} = 0$ on $\zeta \leq T$. On the event $\{\zeta \leq T + t\} = \{\bar{\zeta} \leq t\}$, we have,

$$
Z_{T+t} - Z_T = Z_\zeta + W_{T+t} - W_\zeta
$$
$$
= Z_{\bar{\zeta}} + \tilde{W}_t - \tilde{W}_{\bar{\zeta}}. \tag{5.16}
$$

Equations (5.15) and (5.16) show that $Z_{T+t} - Z_T$ is a solution to (5.14) with the initial condition $U(u, 0, T)$. By uniqueness of solutions we therefore have

$$
Z_{T+t} - Z_T \equiv Z(u, 0, T + t) - Z(u, 0, T) = Z(U(u, 0, T), T, t).
$$

It follows that $\mathbb{P}$-a.s., for all $t \geq 0$,

$$
U(u, 0, T + t) = U(U(u, 0, T), T, t).
$$

We will use the following standard fact. If $\Phi(h, \cdot)$ is independent of the σ-field $\mathcal{H}$ for every h, and H is $\mathcal{H}$-measurable then $\mathbb{E}(\Phi(H, \cdot)|\mathcal{H}) = \phi(H)$ a.s., where $\phi(h) = \mathbb{E}(\Phi(h, \cdot))$. Hence, for any bounded Borel f,

$$
\mathbb{E}\{f(U(u, 0, T + t)) \mid \mathcal{F}_T\} = \mathbb{E}\{f(U(U(u, 0, T), T, t)) \mid \mathcal{F}_T\}
$$
$$
= g(U(u, 0, T)),
$$

where $g(r) = \mathbb{E} f(U(r, T, t))$. However, by pathwise uniqueness of solutions to (5.14), under $\mathbb{P}$, $Z(u, T, \cdot)$ and $Z(u, 0, \cdot)$ are equal in law, and therefore so are $U(u, T, \cdot)$ and $U(u, 0, \cdot)$. Consequently, $g(r) = \mathbb{E} f(U(r, 0, t)) = P_t(r, f)$. This shows that

$$
\mathbb{E}\{f(U(u, 0, T + t)) \mid \mathcal{F}_T\} = P_t(U(u, 0, T), f).
$$

$\square$

Next we will show that the "mirror coupling" (X, Y) has in fact the mirror property.

Proposition 5.7. *Let (X, Y) be a mirror coupling of reflected Brownian motions, A be an event in $\mathcal{F}$, and σ, τ be two a.s. finite $\mathcal{F}$-measurable times with $\sigma \leq \tau$ a.s. on A, such that $X_s, Y_s \notin \partial D$ and $X_s \neq Y_s$ for all $s \in [\sigma, \tau)$ a.s. on A. Then a.s. on A,*

$$(X_s + Y_s) \cdot (Y_\sigma - X_\sigma) = (X_\sigma + Y_\sigma) \cdot (Y_\sigma - X_\sigma), \quad \text{for all } s \in [\sigma, \tau]. \quad (5.17)$$

Proof. Recall that in our notation, $dX = dW + dL$, $dY = dZ + dM$, L and M are the boundary terms for X and Y, and $dZ = H(m)dW = dW - 2m\, m \cdot dW$. Here $m_s = V_s/|V_s|$ on $\{s < \zeta\}$, where $V_s = Y_s - X_s$. Applying Itô's formula, one finds that

$$dm_s = |V_s|^{-1}(dM_s - dL_s) - |V_s|^{-1}m_s m_s \cdot (dM_s - dL_s). \quad (5.18)$$

This shows that the mirror does not move within $[\sigma, \tau]$ on A. Hence for $s \in [\sigma, \tau]$,

$$Z_s = Z_\sigma + G(Y_\sigma - X_\sigma)(W_s - W_\sigma).$$

Since for $s \in [\sigma, \tau]$,

$$X_s = X_\sigma + W_s - W_\sigma, \quad Y_s = Y_\sigma + Z_s - Z_\sigma,$$

(5.17) follows. $\square$

Let (e_1, e_2) be the usual orthonormal basis for $\mathbb{R}^2$ and

$$e_1' = (e_1 - e_2)/\sqrt{2}, \quad e_2' = (e_1 + e_2)/\sqrt{2}.$$

For $x, y \in \mathbb{R}^2$, let $x \leq y$ mean

$$x \cdot e_i' \leq y \cdot e_i', \quad i = 1, 2.$$

Proposition 5.8. *Suppose $D \subset \mathbb{R}^2$ is a piecewise C^2-smooth lip domain for which the defining functions f_1, f_2 (cf. Eq. (5.1)) are both Lipschitz with constants strictly less than one. Let (X, Y) be a mirror coupling of reflected Brownian motions in D, starting from $(x, y) \in (\overline{D})^2$. If $x \leq y$ then $X_t \leq Y_t$ for all $t \geq 0$, a.s.*

Proof. Recall the definition of m_s from the proof of Proposition 5.7. Let a_s be the unit vector perpendicular to m_s such that, in complex number notation, $i a_s = m_s$. Then the identity (5.18) is equivalent to

$$dm_s = |V_s|^{-1}a_s a_s \cdot (dM_s - dL_s).$$

Let $\tau = \inf\{t : X_t \nleq Y_t\}$ and $\Omega_1 = \{\tau < \infty\}$. We will show that Ω_1 has probability 0. Suppose $\tau < \infty$. Since a normally reflected Brownian motion does not hit a fixed point on the boundary, we can assume that X_τ and Y_τ are not at any of the vertices of ∂D. Let $\partial_+ D$ $[\partial_- D]$ denote the intersection of the boundary ∂D with the graph of f_2 [respectively, f_1]. Because the Lipschitz constants of f_1 and f_2 are assumed to be less than 1, it cannot happen that both X_τ and Y_τ are on the same side $\partial_+ D$ or $\partial_- D$ of the boundary. Assume without loss of generality that $Y_\tau \in \partial_+ D$. Then $X_\tau \in D \cup \partial_- D$, $m_\tau = e_2'$ and $a_\tau = e_1'$. It follows from the definition of τ that for every $\varepsilon > 0$ there exists $t \in (\tau, \tau + \varepsilon)$ with $m_t \cdot e_1' < 0$. Since a_t is continuous, we can find small $\varepsilon_0 > 0$ such that for $s \in (\tau, \tau + \varepsilon_0]$, (i) $a_s \cdot e_1' \geq 0$; (ii) $a_s \cdot \mathbf{n}(Y_s) \geq 0$ if $Y_s \in \partial D$; and (iii) $a_s \cdot \mathbf{n}(X_s) \leq 0$ if $X_s \in \partial D$. In (ii) and (iii) we have used the fact that $Y_\tau \in \partial_+ D$ and $X_\tau \notin \partial_+ D$, and the assumption that the Lipschitz constants of f_k's are less than one. For all $t \in (\tau, \tau + \varepsilon_0]$,

$$m_t \cdot e_1' = m_\tau \cdot e_1' + \int_\tau^t |V_s|^{-1} a_s \cdot e_1' \, a_s \cdot (dM_s - dL_s)$$

$$= \int_\tau^t |V_s|^{-1} a_s \cdot e_1' \, a_s \cdot \mathbf{n}(Y_s) d|M|_s - \int_\tau^t |V_s|^{-1} a_s \cdot e_1' \, a_s \cdot \mathbf{n}(X_s) d|L|_s.$$

The first integral on the last line is non-negative and the second one is non-positive. This contradicts the fact that $m_t \cdot e_1' < 0$ for some $t \in (\tau, \tau + \varepsilon_0)$ and so it shows that Ω_1 has probability zero.   □

5.5   Applications of Synchronous and Mirror Couplings to the Hot Spots Problem

This section presents material from [BB99].

5.5.1   Results Based on "Synchronous Couplings"

In this section, we will choose the value of the angle $\angle K$ formed by a straight line K with the horizontal axis so that $\angle K \in (-\pi/2, \pi/2]$. Let $\angle \nabla_x u(t, x)$ be the angle formed by the gradient $\nabla_x u(t, x)$ with the horizontal axis. Recall the definition of a "lip domain" from Sect. 3.1.

Theorem 5.9. *Suppose that D is a lip domain and the Lipschitz constant for the Lipschitz functions in (5.1) is strictly less than 1. Suppose that Γ is a piecewise smooth curve which divides D into two subsets and such that for any line K_1 orthogonal to Γ at a point where Γ is smooth, we have $\angle K_1 \in (-\pi/4, \pi/4)$. Suppose that $u(t, x)$ is a solution to (3.1). Let A be the set to the right of Γ and let $u(0, x) = \mathbf{1}_A(x)$ for $x \in D$. Then for every t and x we have*

$$-\pi/4 \leq \angle \nabla_x u(t, x) \leq \pi/4. \tag{5.19}$$

Sketch of the Proof. Suppose that $x, y \in D$, $x = (x^1, x^2)$, $y = (y^1, y^2)$, and $x^1 < y^1$. Let X_t and Y_t be a pair of reflected Brownian motions in D with $X_0 = x$ and $Y_0 = y$. We assume that (X_t, Y_t) is a synchronous coupling as in (5.2). Let K_t be the line passing through X_t and Y_t. Assume that x and y are such that $-\pi/4 \le \angle K_0 \le \pi/4$. For geometric reasons, $\angle K_t$ can never leave the interval $[-\pi/4, \pi/4]$. Moreover, we will always have $X_t^1 < Y_t^1$. The last two observations and the assumptions on Γ imply that $u(0, X_t) < u(0, Y_t)$ for all t, a.s.

The function $u(t, x)$ may be probabilistically represented as $u(t, x) = \mathbb{E}\, u(0, X_t)$ and, by analogy, $u(t, y) = \mathbb{E}\, u(0, Y_t)$. This and the inequality $u(0, X_t) < u(0, Y_t)$ imply that $u(t, x) = u(t, (x^1, x^2))$ is an increasing function of x^1 for $(x^1, x^2) \in K \cap D$, where K is any line with $-\pi/4 \le \angle K \le \pi/4$. Hence, the gradient $\nabla_x u(t, x)$ must satisfy (5.19). $\square$

Theorem 5.10. *For every lip domain D there exists an eigenfunction $\varphi_2(x)$ corresponding to the second eigenvalue μ_2 for the Neumann problem in D such that for every $y \in D$ we have*

$$\inf_{x \in \partial D} \varphi_2(x) < \varphi_2(y) < \sup_{x \in \partial D} \varphi_2(x).$$

Proof. Fix some eigenfunction $\varphi_2^*(x)$ corresponding to the second eigenvalue μ_2 and find $x_* \in D$ such that $\varphi_2^*(x_*) > 0$. We will consider two curves Γ and Γ_1 and the corresponding regions A and A_1 to the right of Γ and Γ_1, as in Theorem 5.9. We let Γ be the vertical line passing through x_*. We choose Γ_1 in such a way that $A \subset A_1$ and $C = A_1 \setminus A$ is a small triangle which contains x_* in its boundary. We make C so small that, by continuity of φ_2^*, we have $\varphi_2^*(x) > 0$ for $x \in C$. We also require that the sides of C are close to vertical so that we can apply Theorem 5.9 with Γ_1 in place of Γ.

We set $u_0(x) = \mathbf{1}_A(x)$ and $u_0^1(x) = \mathbf{1}_{A_1}(x)$ for $x \in D$. Let $u(t, x)$ and $u^1(t, x)$ be the solutions of the Neumann problem in D with initial conditions $u_0(x)$ and $u_0^1(x)$, respectively. By Proposition 4.1,

$$u(t, x) = \alpha_1 + \alpha_2^* \varphi_2^*(x) e^{-\mu_2 t} + \tilde{\alpha}_2 \tilde{\varphi}_2(x) e^{-\mu_2 t} + R(t, x),$$

where $R(t, x)$ converges to 0 faster than $e^{-\mu_2 t}$ as $t \to \infty$. Here $\tilde{\varphi}_2(x)$ is an eigenfunction corresponding to μ_2 which is orthogonal to $\varphi_2^*(x)$. The analogous formula for $u^1(t, x)$ is

$$u^1(t, x) = \beta_1 + \beta_2^* \varphi_2^*(x) e^{-\mu_2 t} + \tilde{\beta}_2 \tilde{\varphi}_2(x) e^{-\mu_2 t} + R^1(t, x).$$

We have

$$\beta_2^* - \alpha_2^* = \int_{A_1 \cap D} \varphi_2^*(x)dx - \int_{A \cap D} \varphi_2^*(x)dx = \int_C \varphi_2^*(x)dx > 0,$$

so at least one of the coefficients α_2^* or β_2^* is non-zero. Let us assume that $\beta_2^* \ne 0$, the other case being analogous. Then

$$u^1(t, x) = \beta_1 + \beta_2 \varphi_2(x) e^{-\mu_2 t} + R^1(t, x), \tag{5.20}$$

where $\varphi_2(x)$ is a second eigenfunction and $\beta_2 \neq 0$.

Suppose that $\beta_2 > 0$; the other case can be dealt with in a similar way. Theorem 5.9 implies that $u^1(t, x)$ is monotone on every horizontal line passing through D, for every fixed t. Without loss of generality, let us assume that $u^1(t, (x^1, x^2))$ is an increasing function of x^1. For every $x \in D$, let $V(x)$ be the set of $y \in D$ such that the angle between the vector $\overrightarrow{x, y}$ and the horizontal axis lies within $(-\pi/4, \pi/4)$. By Theorem 5.9 and our choice of Γ and Γ_1, we have $u^1(t, y) \geq u^1(t, x)$, for all $x \in D$, $y \in V(x)$, and $t > 0$. Since $R^1(t, x)$ converges to 0 faster than $e^{-\mu_2 t}$, the last fact and (5.20) imply that $\varphi_2(y) \geq \varphi_2(x)$ for $y \in V(x)$. The remark following [Fol76, Corollary 6.31] may be applied to the operator $\Delta + \mu_2$ to conclude that the eigenfunctions are real analytic and therefore they cannot be constant on an open set unless they are constant on the whole domain D. It follows that the maximum of $\varphi_2(x)$ cannot be attained on an open subset of D and thus it can be attained only at the right vertex. The proof that the minimum is attained at the left vertex is completely analogous.   □

5.5.2 Results Based on "Mirror Couplings"

We start with a discussion of mirror couplings that is more detailed than that given in Sect. 3.3 but less technical and more relevant for the present section than the construction and properties presented in Sect. 5.1.

First we will describe the mirror coupling for free Brownian motions in $\mathbb{R}^2$. Suppose that $x, y \in \mathbb{R}^2$, $x \neq y$, and that x and y are symmetric with respect to a line M. Let X_t be a Brownian motion starting from x and let τ be the first time t with $X_t \in M$. Then we let Y_t be the mirror image of X_t with respect to M for $t \leq \tau$, and we let $Y_t = X_t$ for $t > \tau$. The process Y_t is a Brownian motion starting from y. The pair (X_t, Y_t) is a "mirror coupling" of Brownian motions in $\mathbb{R}^2$.

Next we turn to the mirror coupling of reflected Brownian motions in a half-plane $\mathcal{H}$, starting from $x, y \in \mathcal{H}$. Let M be the line of symmetry for x and y. The case when M is parallel to $\partial \mathcal{H}$ can be easily handled using Skorohod's lemma, so we focus on the case when M intersects $\partial \mathcal{H}$. By performing rotation and translation, if necessary, we may suppose that $\mathcal{H}$ is the upper half-plane and M passes through the origin. We will write $x = (r^x, \theta^x)$ and $y = (r^y, \theta^y)$ in the polar coordinates. The points x and y are at the same distance from the origin so $r^x = r^y$. Suppose without loss of generality that $\theta^x < \theta^y$. We first generate a two-dimensional Bessel process R_t starting from r^x. Then we generate two coupled one-dimensional processes on the "half-circle" as follows. Let $\tilde{\Theta}^x_t$ be a one-dimensional Brownian motion starting from θ^x. Let $\tilde{\Theta}^y_t = -\tilde{\Theta}^x_t + \theta^x + \theta^y$. Let Θ^x_t be the reflected Brownian motion on $[0, \pi]$, constructed from $\tilde{\Theta}^x_t$ by the means of the Skorohod lemma, using "local time" push on both sides of the interval $[0, \pi]$. The analogous reflected process obtained from $\tilde{\Theta}_t^y$ will be denoted $\hat{\Theta}_t^y$. Let τ^Θ be the smallest t with $\Theta_t^x = \hat{\Theta}_t^y$. Then we let $\Theta_t^y = \hat{\Theta}_t^y$ for $t \leq \tau^\Theta$ and $\Theta_t^y = \Theta_t^x$ for $t > \tau^\Theta$. We define a "clock" by $\sigma(t) = \int_0^t R_s^{-2} ds$. Then $X_t = (R_t, \Theta_{\sigma(t)}^x)$ and $Y_t = (R_t, \Theta_{\sigma(t)}^y)$ are reflected Brownian motions in $\mathcal{H}$ with the normal vector of reflection. Moreover, X_t and Y_t behave like free Brownian motions coupled by the mirror coupling as long as they are both strictly inside $\mathcal{H}$. The processes will stay together after the first time they meet. This property is crucial in this section but was hardly relevant for the synchronous coupling. For definiteness, we let M_t be the horizontal line passing through X_t if $X_t = Y_t$.

Homework 5.11. Give a formal proof of the assertions stated in the last paragraph based on the "skew product" decomposition; see [IM74, Sect. 7.15].

The most important property of the above coupling is that the two processes X_t and Y_t remain at the same distance from a fixed point (the origin). We will describe how this property manifests itself in more general settings. First of all, suppose that $\mathcal{H}$ is again an arbitrary half-plane, and x and y belong to $\mathcal{H}$. Let M be the line of symmetry for x and y. Then our construction generates a pair of reflected Brownian motions starting from x and y such that the distance from X_t to $M \cap \partial\mathcal{H}$ is the same as for Y_t, for every t. Let M_t be the line of symmetry for X_t and Y_t. Note that M_t may move, but only in a continuous way, while the point $M_t \cap \partial\mathcal{H}$ will never move. We will call M_t the *mirror* and the point $H = M_t \cap \partial\mathcal{H}$ will be called the *hinge*. The absolute value of the angle between the mirror and the normal vector to $\partial\mathcal{H}$ at H can only decrease.

The mirror coupling of reflected Brownian motions in a convex polygonal domain D can be described as follows. Suppose that X_t and Y_t start from x and y inside the domain D. As soon as one of the particles hits a side I of ∂D, the processes will evolve according to the coupling described in the previous paragraph. To be more precise, let K be the straight line containing I. Since the process which hits I does not "feel" the shape of D except for the direction of I, it follows that the two processes will remain at the same distance from the hinge $H_t = M_t \cap K$. The mirror M_t can move but the hinge H_t will remain constant as long as I remains the side of ∂D where the reflection takes place. The hinge H_t will jump when the reflection location moves from I to another side of ∂D. Since D is convex, H_t will be always on ∂D or outside D.

We will say that X_t is *active* if it is currently reflecting from a side of ∂D and similarly for Y_t. Let U_t^1 and U_t^2 be the intersection points of the mirror M_t with ∂D. Let ∂D^a be the "active" part of ∂D, i.e., this connected component of $\partial D \setminus \{U_t^1, U_t^2\}$ which contains the active particle. We note that the active part ∂D^a can only increase with time as a subset of the boundary. However the active part will switch from one side of M_t to the other from time to time. It will later turn out that this is a convenient way to describe all possible movements of the mirror M_t.

Theorem 5.12. *Suppose that D is a convex planar polygonal domain which is symmetric with respect to the horizontal axis. Let z_L and z_R be the intersection points of ∂D with the horizontal axis. Assume that for every $r > 0$, the intersection*

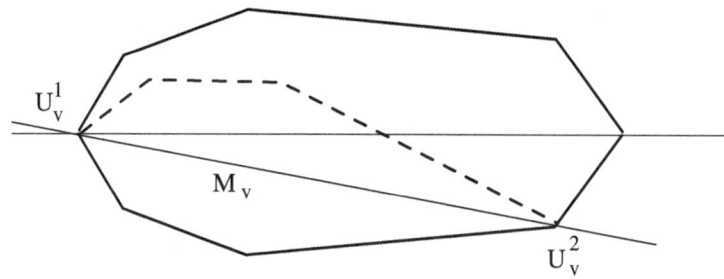

Fig. 5.1 The *dotted line* is the mirror image, with respect to M_v, of the non-active side of the boundary just before time v

of the circle $\partial\mathcal{B}(z_L, r)$ with D is either empty or it is a connected arc, and the same holds for $\partial\mathcal{B}(z_R, r)$. Let Γ be any vertical line and let A be the half-plane to the right of Γ. Consider the solution $u(t, x)$ to the heat equation with Neumann boundary conditions in D and with the initial condition $u(0, x) = \mathbf{1}_A(x)$ for $x \in D$. Suppose that $x_0 \in D$ lies above the horizontal axis. Then for every t, the line containing $\nabla_x u(t, x_0)$ and passing through x_0 passes above or through each of the points z_L and z_R. An analogous statement holds for points on the other side of the horizontal axis, by symmetry. The horizontal component of $\nabla_x u(t, x_0)$ points to the right, for every $x_0 \in D$ and t.

Note that every ellipse D can be oriented in such a way that it satisfies the condition that for every $r > 0$, the intersection of the circle $\partial\mathcal{B}(z_L, r)$ with D is either empty or it is a connected arc, and the same holds for $\partial\mathcal{B}(z_R, r)$.

Proof. Consider any straight line M which intersects the upper part of ∂D at a point U^1 and the lower part of ∂D at U^2. Take any points $x, y \in D$ which are symmetric with respect to M and let X_t and Y_t be reflected Brownian motions in D, starting from x and y, respectively, and related by the mirror coupling. Recall that the mirror for X_t and Y_t is denoted M_t. As long as M_t intersects both the upper and the lower parts of ∂D, we will denote the intersection points U_t^1 and U_t^2. This is true for $t = 0$, by assumption, and we intend to prove that this will remain true for all t, a.s.

The points U_t^1 and U_t^2 move in a continuous fashion because M_t does. Recall that the active part ∂D^a of the boundary can only increase. This means that both U_t^1 and U_t^2 move towards z_L or both points move towards z_R. Both points can reach either one of these points at the same time only if X_t and Y_t hit one of these points. This event has probability zero by a theorem in [VW85]. Suppose that U_t^1 reaches z_L at time v, before U_t^2 does. Then M_v forms a negative angle with the horizontal axis. Since the points U_t^1 and U_t^2 must have been moving towards z_L just before the time v, it follows that the active side ∂D^a of the boundary was above and to the right of M_t. Figure 5.1 illustrates the fact that the mirror image of the non-active side of the boundary with respect to M_v lies strictly inside D—this is due to the assumption that for every $r > 0$, the intersection of the circle $\partial\mathcal{B}(z_L, r)$ with D is either empty or it is a connected arc, and the same holds for $\partial\mathcal{B}(z_R, r)$. We obtain a contradiction as both processes X_t and Y_t must always stay within the set $\overline{D}$ and they are always mirror images of each other with respect to M_t (or $X_t = Y_t$).

The same argument shows that none of the points U_t^1 or U_t^2 can hit z_L or z_R before the coupling time for X_t and Y_t. This implies that the mirror M_t cannot attain the horizontal direction before the coupling time and so we conclude that $X_t^1 - Y_t^1$ does not change the sign. This implies, in the same way as in the proof of Theorem 5.9, that for every t, the function $u(t,x) = u(t,(x^1,x^2))$ is increasing in x^1 on every straight line M_1 which is perpendicular to any line M which crosses the upper and lower parts of ∂D. This easily implies the claim about the direction of the gradient $\nabla_x u(t,x_0)$.   □

Recall that we say that a function $h(x^1,x^2)$ is antisymmetric with respect to the horizontal axis if $h(x^1,x^2) = -h(x^1,-x^2)$ for all (x^1,x^2).

Theorem 5.13. *Suppose that D is a convex polygonal planar domain satisfying the hypotheses of Theorem 5.12 and there is no eigenfunction $\varphi_2(x)$ corresponding to the second eigenvalue which is antisymmetric with respect to the horizontal axis. Then there exists an eigenfunction $\varphi_2(x)$ corresponding to the second eigenvalue such that for every $y \in D$ we have $\inf_{x \in \partial D} \varphi_2(x) < \varphi_2(y) < \sup_{x \in \partial D} \varphi_2(x)$.*

Proof. Let z_L and z_R be intersection points of ∂D with the horizontal axis. Take any eigenfunction $\varphi_2^*(x)$ corresponding to the second eigenvalue and let $\varphi_2(x^1,x^2) = \varphi_2^*(x^1,x^2) + \varphi_2^*(x^1,-x^2)$ for all $(x^1,x^2) \in D$. Note that $\varphi_2(x)$ is an eigenfunction corresponding to the second eigenvalue. By assumption, $\varphi_2(x)$ is not identically equal to zero. By Courant's nodal domain theorem ([Cha84, p. 19] or [Ban80, p. 112]), the nodal line of $\varphi_2(x)$ divides D into only two nodal domains. This and the fact that $\varphi_2(x)$ is symmetric with respect to the horizontal axis imply that the nodal line must lie at a positive distance from the points z_L and z_R. Hence, $\varphi_2(x)$ does not vanish at either point and this is also true for some neighborhoods of both points, by the continuity of $\varphi_2(x)$. We will suppose that $\varphi_2(z_R) > 0$; the proof is analogous when we have the opposite inequality.

The rest of the proof is very similar to the proof of Theorem 5.10. Consider two distinct vertical lines Γ and Γ_1 and the corresponding regions A and A_1 to the right of Γ and Γ_1. We assume the Γ and Γ_1 are so close to z_R that $\varphi_2(x) > 0$ for all $x \in (A \cup A_1) \cap D$.

We set $u_0(x) = \mathbf{1}_A(x)$ and $u_0^1(x) = \mathbf{1}_{A_1}(x)$ for $x \in D$. Let $u(t,x)$ and $u^1(t,x)$ be the solutions of the Neumann problem in D with initial conditions $u_0(x)$ and $u_0^1(x)$, respectively. We have, by Proposition 4.1,

$$u(t,x) = \alpha_1 + \alpha_2 \varphi_2(x) e^{-\mu_2 t} + \tilde{\alpha}_2 \tilde{\varphi}_2(x) e^{-\mu_2 t} + R(t,x),$$

where $R(t,x)$ converges to 0 faster than $e^{-\mu_2 t}$ as $t \to \infty$. Here $\tilde{\varphi}_2(x)$ is an eigenfunction corresponding to μ_2 which is orthogonal to $\varphi_2(x)$. The analogous formula for $u^1(t,x)$ is

$$u^1(t,x) = \beta_1 + \beta_2 \varphi_2(x) e^{-\mu_2 t} + \tilde{\beta}_2 \tilde{\varphi}_2(x) e^{-\mu_2 t} + R^1(t,x).$$

We have

$$\beta_2 - \alpha_2 = \int_{A_1 \cap D} \varphi_2(x)dx - \int_{A \cap D} \varphi_2(x)dx = \int_{(A_1 \setminus A) \cap D} \varphi_2(x)dx \neq 0,$$

so at least one of the coefficients α_2 or β_2 is non-zero. Let us assume that $\beta_2 \neq 0$, the other case being analogous. Then

$$u^1(t, x) = \beta_1 + \hat{\beta}_2 \hat{\varphi}_2(x)e^{-\mu_2 t} + R^1(t, x), \tag{5.21}$$

where $\hat{\varphi}_2(x)$ is a second eigenfunction and $\hat{\beta}_2 \neq 0$.

Without loss of generality we assume that $\hat{\beta}_2 > 0$. Theorem 5.12 implies that $u^1(t, (x^1, x^2))$ is an increasing function of x^1, for every fixed t. For every $x \in D$, let $V(x)$ be the intersection of D with the ball $\mathcal{B}(z_R, |x - z_R|)$. Using the information about the direction of the gradient $\nabla_x u(t, x)$ provided by Theorem 5.12, it is elementary to see that $u^1(t, y) \geq u^1(t, x)$, for all $x \in D$, $y \in V(x)$, and $t > 0$. Since $R^1(t, x)$ converges to 0 faster than $e^{-\mu_2 t}$, the last fact and (5.21) imply that $\hat{\varphi}_2(y) \geq \hat{\varphi}_2(x)$ for $y \in V(x)$. Since the maximum of $\hat{\varphi}_2(x)$ cannot be attained on a non-empty open subset of D (recall the argument from the proof of Theorem 5.10), it can be attained only at z_R. The location of the minimum is z_L, by the same argument. □

Theorem 5.14. *Suppose that D is a convex polygonal planar domain which is symmetric with respect to the horizontal and vertical axes. Then there exists an eigenfunction $\varphi_2(x)$ corresponding to the second eigenvalue such that for every $y \in D$ we have $\inf_{x \in \partial D} \varphi_2(x) < \varphi_2(y) < \sup_{x \in \partial D} \varphi_2(x)$.*

Proof. First we are going to show that there exists an eigenfunction $\varphi_2(x)$ which is antisymmetric with respect to one of the axes, i.e., we either have $\varphi_2(x^1, x^2) = -\varphi_2(x^1, -x^2)$ for all $(x^1, x^2) \in D$ or we have $\varphi_2(x^1, x^2) = -\varphi_2(-x^1, x^2)$ for all $(x^1, x^2) \in D$. Take any eigenfunction $\varphi_2^*(x)$ corresponding to the second eigenvalue. Let $\tilde{\varphi}_2(x^1, x^2) = \varphi_2^*(x^1, x^2) + \varphi_2^*(-x^1, x^2)$. If $\tilde{\varphi}_2(x)$ is identically 0 then we can take $\varphi_2(x) = \varphi_2^*(x)$. Otherwise we let $\hat{\varphi}_2(x^1, x^2) = \tilde{\varphi}_2(x^1, x^2) + \tilde{\varphi}_2(x^1, -x^2)$.

We will prove that $\hat{\varphi}_2(x)$ is identically equal to 0. Suppose that $\hat{\varphi}_2(x)$ is not identically equal to 0. The function $\hat{\varphi}_2(x)$ is symmetric with respect to both axes. Let $\tilde{D}$ be the part of D in the first quadrant. Since $\hat{\varphi}_2(x)$ must take both positive and negative values and it is symmetric with respect to both axes, the nodal line must intersect the interior of $\tilde{D}$. The nodal line cannot form a closed loop inside $\tilde{D}$, see, e.g., [Ban80, p. 128]. (The fact that the nodal line cannot form a closed loop follows easily from the fact that $\lambda_1 > \mu_2$, [Ban80, p. 155], where λ_1 is first Dirichlet eigenvalue for D.) The part of the nodal line inside $\tilde{D}$ cannot touch both axes because, by symmetry, we would have a closed loop formed by the nodal line inside D. If the nodal line inside $\tilde{D}$ touches ∂D, then D must be divided into more than two nodal domains. This is ruled out by the Courant nodal domain theorem quoted in the proof of Theorem 5.13. This completes the proof that $\hat{\varphi}_2(x)$ is identically equal to 0 and so we can take $\varphi_2(x) = \tilde{\varphi}_2(x)$.

The nodal line for $\varphi_2(x)$ must lie on one of the axes. Without loss of generality, suppose that it lies on the vertical axis. Let $\mathcal{H}^+$ denote the right half plane. Then $\varphi_2(x)$ is the first eigenfunction for the mixed Dirichlet-Neumann problem in $D' = D \cap \mathcal{H}^+$, with the Dirichlet boundary conditions on $\partial_1 D' = \partial D \cap \partial \mathcal{H}^+$ and the Neumann boundary conditions elsewhere on the boundary. We will prove that $\varphi_2(x)$ is monotone on all horizontal lines.

The probabilistic representation of the solutions $u(t, x)$ to the Dirichlet-Neumann heat problem involves Brownian motion X_t reflected on $\partial_2 D' = \partial D \cap \mathcal{H}^+$ and killed on $\partial_1 D'$. Let $u_0(x) = 1$ for all $x \in D'$. If $u(0, x) = u_0(x)$ for all $x \in D'$ and $X_0 = x$ then $u(s, x)$ is equal to the probability that X_t is not killed on $\partial_1 D'$ before time s. Suppose that $x = (x^1, x^2)$ and $y = (y^1, y^2)$ are any points in D' with $x^2 = y^2$, and $x^1 < y^1$. In order to prove monotonicity of $\varphi_2(x)$ on horizontal lines it will suffice to construct Brownian motions X_t and Y_t, starting from x and y, and such that X_t exits D' through $\partial_1 D'$ no later than Y_t does.

Our proof will use the mirror coupling except that if any of the processes X_t or Y_t hits $\partial_1 D'$, it will be killed, and the other process, if it survives beyond this point, will continue on its own. The other process may be killed later.

Since the points x and y lie on a horizontal line, the initial direction of the mirror M_0 for X_0 and Y_0 is vertical. Let U_0^1 and U_0^2 be the upper and lower points of intersection of M_0 with $\partial D'$. Since M_t moves in a continuous way, we can choose the labels U_t^1 and U_t^2 for the intersection points of M_t with $\partial D'$ in such a way that U_t^1 and U_t^2 are continuous functions of t. In this proof we change the conventions concerning the angles and we choose the angle $\angle M_t$ between M_t and the horizontal axis so that $t \to \angle M_t$ is a continuous function. We set $\angle M_0 = \pi/2$.

Let z_R be the intersection point of $\partial_2 D'$ and the horizontal axis. We will argue that neither U_t^1 nor U_t^2 can ever touch z_R and $\angle M_t$ always stays in $[0, \pi]$. Suppose that this is not always true and let v be the infimum of t such that $U_t^1 = z_R$ or $U_t^2 = z_R$ or $\angle M_t \notin [0, \pi]$. First we consider the case when $U_v^1 = z_R$ (the case $U_v^2 = z_R$ is analogous). One can prove that M_v cannot be horizontal in this case but we do not need to do this—if M_v is horizontal and $U_v^1 = z_R$ then M_v lies on the horizontal axis, so, by symmetry, the line M_t will stop moving at time v and X_t and Y_t will hit $\partial_1 D'$ at the same time.

Next suppose that $U_v^1 = z_R$ and M_v is not horizontal, and so $\angle M_v \in (0, \pi/2)$. In this case, the argument is very similar to that in the proof of Theorem 5.12. The main difference is that we do not assume any more that for every $r > 0$, the intersection of the circle $\partial \mathcal{B}(z_L, r)$ with D is either empty or it is a connected arc, and the same holds for $\partial \mathcal{B}(z_R, r)$. However, we use the assumption of D having two lines of symmetry as follows. Note that the active part of the boundary must have been the part of $\partial_2 D'$ above U_t^1, just before time v, because U_t^1 was pushed down to z_R. The mirror image, with respect to M_v, of the part of $\partial_2 D'$ below M_v lies strictly inside D', or on $\partial_1 D'$, or outside D'. This contradicts the fact that X_t and Y_t must always stay inside the domain D', and that the active side of the boundary just before time v was above U_t^1.

Now suppose that $\angle M_v = 0$; as usual, the symmetric case $\angle M_v = \pi$, is left to the reader. We have already discussed the case when M_v lies on the horizontal axis, so let us assume that it does not. Suppose U_v^1 lies in the first quadrant. Then the mirror image, with respect to M_v, of the part of D' above M_v lies inside the part of D' below M_v. We have two possibilities for what may happen after time v. First suppose that there is a vertical line segment on the boundary of D' and z_R lies on this vertical segment. Then both X_t and Y_t may be reflecting at the same time from this line segment for some time after v. In this case M_t will not be moving.

The only other possibility is that the upper side of the boundary of D' will be active. If a part of the boundary of D' is horizontal and one of the processes is reflecting from this part, the mirror will move but it will not change its direction. Otherwise, since D is symmetric with respect to the vertical axis, the hinge for the mirror, if it exists, must lie to the right of $M_v \cap \partial_2 D'$ and so the mirror will be turning counterclockwise, i.e., the angle $\angle M_t$ will move from 0 to inside the interval $(0, \pi/2)$.

It is routine to restart the argument at the next time when $\angle M_v = 0$ or $\angle M_v = \pi$ and complete the proof of the claim that U_t^1 and U_t^2 never hit z_R and $\angle M_t$ always stays in $[0, \pi]$.

Since $\angle M_t \in [0, \pi]$ for all t, we have $X_t^1 \le Y_t^1$ for all t, until one or both of the processes are killed. This proves that X_t must hit $\partial_1 D'$ before or at the same time when Y_t hits $\partial_1 D'$. This in turn proves the monotonicity of $u(t, x)$ along horizontal lines within D', for every fixed t.

Next we extend our argument to points $x = (x^1, x^2)$ and $y = (y^1, y^2)$ such that the line of symmetry for these points crosses both the upper and the lower sides of ∂D. The same reasoning as for x and y lying on a horizontal line proves that if $x^1 < y^1$ then the process starting from x will hit $\partial_1 D'$ no later than the process starting from y. It follows that $u(t, x) \le u(t, y)$ for all t. We recall from the proof of Theorem 5.12 that this implies that given any $x_0 \in D$ which lies above the horizontal axis and any t, the line containing $\nabla_x u(t, x_0)$ and passing through x_0 passes above or through each of the points z_L and z_R. Then we can argue as in the proof of Theorem 5.13 that if $V(x)$ is the intersection of D with the ball $\mathcal{B}(z_R, |x - z_R|)$ then $u(t, x) \le u(t, y)$, for all $y \in V(x)$ and all t.

We have $u(t, x) = \alpha \varphi_2(x) e^{-\mu_2 t} + R(t, x)$, where $\alpha \ne 0$ and $R(t, x)$ goes to 0 faster than $e^{-\mu_2 t}$. Without loss of generality suppose that $\alpha > 0$. It follows that $\varphi_2(x) \le \varphi_2(y)$ for $y \in V(x)$. This implies that $\varphi_2(x)$ attains its maximum only at z_R. For the same reason, the minimum is attained at z_L.   □

Homework 5.15. Let Γ_a^1 be a curve defined by parametric equations as

$$\Gamma_a^1 = \{(x^1(s), x^2(s)), 0 \le s \le 2\pi\},$$

$$x^1(s) = -1 + \left(1 + \frac{a}{2\pi}s\right) \frac{s + 2\pi}{4\pi} \cos s,$$

$$x^2(s) = \left(1 + \frac{a}{2\pi}s\right) \frac{s + 2\pi}{4\pi} \sin s.$$

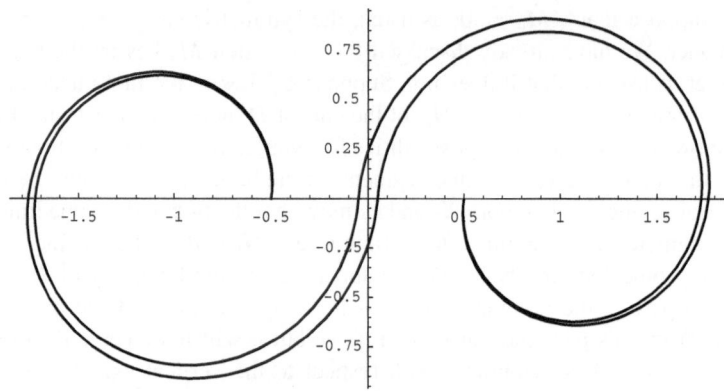

Fig. 5.2 Domain with "monotone" second Neumann eigenfunction

Let Γ_a be the union of Γ_a^1 and the curve symmetric to Γ_{-a}^1 with respect to the point $(0,0)$. The union of $\Gamma_{0.05}$ and $\Gamma_{-0.05}$ forms the boundary of a domain D depicted in Fig. 5.2. Suppose that the function $u(0,x)$ (i.e., the initial condition in (3.1)) is equal to 1 for all $x = (x^1, x^2) \in D$ with $x^1 > -1/4$, $x^2 > 0$, and also for all points with $x^1 > 1/4$. The initial condition is zero elsewhere in D. Prove that for every fixed $t > 0$, the solution $u(x,t)$ of (3.1) is monotone along every line Γ_a, for every $a \in (-0.05, 0.05)$. Hint: The synchronous coupling does not work in this case but the mirror coupling does.

5.6   Hot Spots in a Domain with One Hole

This section is based on [Bur05]. Let D be the planar domain defined in Sect. 4.1 and let $D_1 = \{(x_1, x_2) \in D : x_1 > 0, x_2 > -1\}$. See Fig. 4.2.

Let φ be the second Neumann eigenfunction in D,

$$a = \sup_{(x_1,x_2)\in D_1, x_1=1} \varphi(x),$$

$$\Lambda = \{x \in D_1 : \varphi(x) = a\},$$

$$r = \inf_{(x_1,x_2)\in\Lambda} x_1.$$

It was proved in [Bur05] that for small ε we have

$$r \geq 1 - 5\varepsilon,$$

$$\inf_{(x_1,x_2)\in\overline{D}_1, x_1\leq 1/2} \varphi(x) > \sup_{(x_1,x_2)\in\overline{D}_1, x_1\geq 1-5\varepsilon} \varphi(x).$$

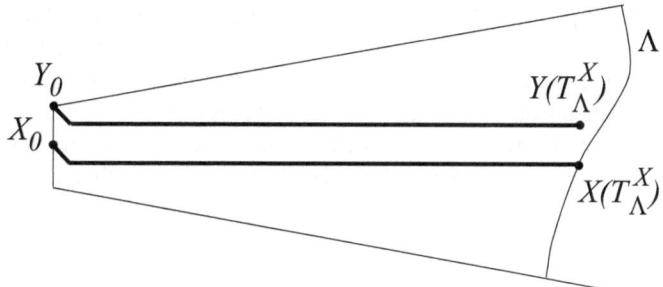

Fig. 5.3 A cartoon of non-coupling Brownian paths traveling far in an "unlikely" manner

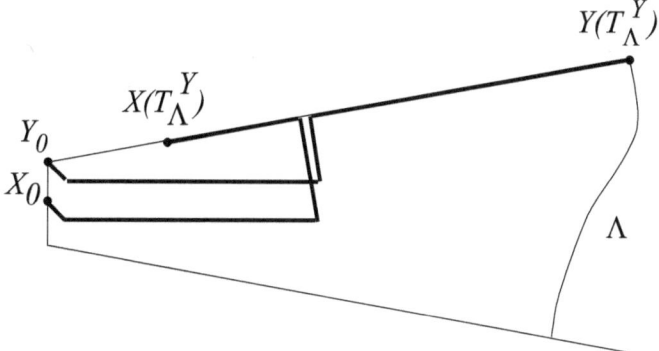

Fig. 5.4 A cartoon of non-coupling Brownian paths traveling far in a desirable manner

We will now define a coupling (X, Y) of reflected Brownian motions in D_1 with $X_0 = (0, 0)$ and $Y_0 = (0, \varepsilon)$. We will give only an intuitive overview, accompanied by pictures; see [Bur05] for detailed construction. Brownian paths will be represented as thick polygonal lines in the figures.

We will construct the coupling (X, Y) so that the event $\{T_\Lambda^X < T_\Lambda^Y\}$ is very unlikely. More precisely, its probability is less than $e^{-c_1/\varepsilon}$ for some constant c_1. The processes are defined in such a way that this event can happen only if X and Y travel in an almost parallel way, as in Fig. 5.3. For the event depicted in Fig. 5.3 to happen, the particles have to avoid coupling when they travel towards Λ, and for some $p_1 > 0$, they can couple on every interval of length ε with probability greater than p_1. An upper estimate for the probability of $\{T_\Lambda^X < T_\Lambda^Y\}$ is $(1-p_1)^{1/\varepsilon} = e^{-c_1/\varepsilon}$.

We need the coupling (X, Y) to have the property that the event $\{T_\Lambda^X > T_\Lambda^Y\}$ is much more likely than $\{T_\Lambda^X < T_\Lambda^Y\}$, and, moreover, $X(T_\Lambda^Y)$ is very far to the left from Λ. To achieve this goal, we construct (X, Y) so that the particles travel to the right in a parallel fashion, then they hit the upper part of the boundary of D_1, and then they travel in the opposite directions (see Fig. 5.4). Moving in parallel to the right is very unlikely, but since the particles have to cover only part of the distance from the starting point to Λ, this event has probability bounded below by $e^{-c_2/\varepsilon}$, where $0 < c_2 < c_1$. For small $\varepsilon > 0$, $e^{-c_2/\varepsilon}$ is much larger than the probability of the event in Fig. 5.3, i.e., $e^{-c_1/\varepsilon}$. The probability that each of the particles follows the last two segments of its path depicted in Fig. 5.4 is only polynomially small in ε.

Finally, the two Brownian particles can follow trajectories completely different from those depicted in Figs. 5.3 and 5.4, but it can be shown that such paths only help to increase the probability of $\{T_\Lambda^X > T_\Lambda^Y\}$. In other words, an important feature of the construction is that in all of the remaining cases, $\{T_\Lambda^X < T_\Lambda^Y\}$ cannot happen.

Homework 5.16. Assume that a coupling with the properties described above exists. Prove that $\varphi(0,0) > \varphi(0,\varepsilon)$.

Chapter 6
Parabolic Boundary Harnack Principle

This chapter is based on [BB92]. The following result is a lemma in [BB92] needed to prove a version of "parabolic boundary Harnack principle" reproduced below as Theorem 6.2. Paradoxically, the "lemma" found many more applications in author's research than the "theorem." We will assume that $D \subset \mathbb{R}^d$ is a bounded Lipschitz domain and X is Brownian motion, to keep the presentation simple. The setting is much more general in [BB92].

The hitting time of a set A by the process X will be denoted $T(A)$.

Lemma 6.1. *For every $u > 0$ there exist a non-degenerate closed ball $M \subset D$ and $c > 0$ such that for all $x \in D$,*

$$\mathbb{P}^x(X_u \in M, T(D^c) > u) \geq c\, \mathbb{P}^x(T(D^c) > u).$$

Proof. Let

$$A = A(\beta) = \{x \in D : \operatorname{dist}(x, \partial D) \geq \beta\}.$$

It is easy to see that A is a bounded and closed set, hence, a compact set.

Fix some $z \in D$ and find $\beta_0 > 0$ and a closed ball M such that $M \subset A(\beta_0)$. Let

$$D_1 := D \setminus M,$$

$$h(x) := G_D^L(x, z),$$

$$D_2 = D_2(\beta) := D \setminus A(\beta),$$

$$U_k := \{x \in D_1 : h(x) \in [2^k, 2^{k+1}]\},$$

$$\hat{U}_k := \{x \in D_1 : h(x) \leq 2^{k+1}\},$$

$$\tilde{U}_k := \{x \in D_2 : h(x) \in [2^k, 2^{k+1}]\}.$$

K. Burdzy, *Brownian Motion and its Applications to Mathematical Analysis,*
Lecture Notes in Mathematics 2106, DOI 10.1007/978-3-319-04394-4_6,
© Springer International Publishing Switzerland 2014

We have $\tilde{U}_k = U_k \setminus A(\beta)$ for $\beta < \beta_0$. Note that U_k is bounded, so it has finite volume and, therefore, $\mathrm{vol}(\tilde{U}_k) \to 0$ as $\beta \to 0$. For any open set N and any x we have, by [Bañ87, Lemma 1],

$$\mathbb{E}^x(T(N^c)) \le c_1 (\mathrm{vol}(N))^{1/d}.$$

It follows that

$$\lim_{\beta \to 0} \mathbb{E}^x(T(\tilde{U}_k^c)) = 0. \tag{6.1}$$

The function h is bounded in D_1 by 2^{k_0+1} for some $k_0 < \infty$.
    It can be shown that (see [BB92, Proof of Theorem 1.1 (i)])

$$\mathbb{E}^x(T(U_k^c)) \le c_* |k|^{-1-\delta}$$

for some $c_* < \infty, \delta > 0$ and all x and k. Hence,

$$\sum_{k=-\infty}^{k_0} \sup_{x \in U_k} \mathbb{E}^x(T(U_k^c)) < \infty. \tag{6.2}$$

Since $\tilde{U}_k \subset U_k$,

$$\mathbb{E}^x(T(\tilde{U}_k^c)) \le \mathbb{E}^x(T(U_k^c)).$$

This and (6.1)–(6.2) show that for any constant $c_2 < \infty$ there is $\beta > 0$ with

$$c_2 \sum_{k=-\infty}^{k_0} \sup_{x \in \tilde{U}_k} \mathbb{E}^x(T(\tilde{U}_k^c)) < u/8.$$

For suitable c_2, the expression on the left hand side is an upper bound for $\mathbb{E}_h^x(T(D_2^c))$ where $\mathbb{E}_h^x$ refers to the h-process (see Sect. 1.9.2). The last claim can be found in [BB92, Proof of Theorem 1.1 (i)]; the idea goes back to [Chu84]. It follows that

$$\mathbb{P}_h^x(T(D_2^c)) < u/4) > 1/2. \tag{6.3}$$

Let the $\mathbb{R}^d$-valued Brownian motion be denoted by X and let Y stand for the space-time process. More precisely, if X has law $\mathbb{P}^x$, then the law of the space-time diffusion

$$\{Y(t) := (X(t), s - t), t \ge 0\}$$

will be denoted $\mathbb{P}^{x,s}$. The distribution of space-time Brownian motion conditioned by a parabolic function g will be denoted $\mathbb{P}_g^{x,s}$. By abuse of notation, $T(A)$ will denote the first hitting time of A for Y as well as for X. The function

$$(x,t) \mapsto g(x,t) := \mathbb{P}^x(T(\partial D) > t)$$

is parabolic in $D \times [0,\infty)$ with boundary values 1 on $D \times \{0\}$ and 0 otherwise.

Let g_1 be a parabolic function in $D \times [0,\infty)$ which has the same boundary values as g except that $g_1(x,0) = \varepsilon$ for $x \in D \backslash M$, where $\varepsilon \in (0,1)$ will be chosen later. We will estimate g_1 on $D \times [u/2, u]$.

It is easy to see that $g_1(x,s) > c_3$ for all $x \in M$ and $s \in [u/4, u]$. We also have $h(y) < c_4$ for all $y \in \partial D_2$. Let $h(x,s) := h(x)$. For $x \in D_2$ and $s \geq 1/2$ we have, by (6.3),

$$g_1(x,s) \geq \int_{t \in [u/4,u], y \in \partial D_2} g_1(y,t)\, \mathbb{P}^{x,s}(T(D_2^c) \in dt, X(T(D_2^c)) \in dy)$$

$$= \int_{t \in [u/4,u], y \in \partial D_2} \frac{h(x,s)}{h(y,t)} \frac{h(y,t)}{h(x,s)} g_1(y,t)\, \mathbb{P}^{x,s}(T(D_2^c) \in dt, X(T(D_2^c)) \in dy)$$

$$= \int_{t \in [u/4,u], y \in \partial D_2} \frac{h(x,s)}{h(y,t)} g_1(y,t)\, \mathbb{P}_h^{x,s}(T(D_2^c) \in dt, X(T(D_2^c)) \in dy)$$

$$\geq \int_{t \in [u/4,u], y \in \partial D_2} h(x,s) c_4^{-1} c_3\, \mathbb{P}_h^{x,s}(T(D_2^c) \in dt, X(T(D_2^c)) \in dy)$$

$$= h(x,s) c_4^{-1} c_3\, \mathbb{P}_h^{x,s}(T(D_2^c) \in [u/4, s])$$

$$\geq h(x,s) c_4^{-1} c_3/2$$

$$= c_5 h(x,s) = c_5 h(x).$$

Let

$$W_k = \{(x,s) : g_1(x,s) \in [2^k, 2^{k+1}], s \in [u/2, u]\},$$

$$W = \bigcup_{k=-\infty}^{k_1} W_k,$$

where $k_1 < 0$ will be chosen later. If $2^{-m} < c_5$ then $W_k \subset \hat{U}_{k+m} \times [u/2, u]$. Using an estimate from [Chu84] we obtain for small k_1

$$\mathbb{E}_{g_1}^{x,u}(T(W^c)) \leq c_6 \sum_{k=-\infty}^{k_1} \sup_{(y,s) \in W_k} \mathbb{E}^{y,s}\, T(W_k^c)$$

$$\leq c_6 \sum_{k=-\infty}^{k_1} \sup_{(y,s) \in \hat{U}_{k+m}} \mathbb{E}^{y,s}\, T(\hat{U}_{k+m}^c) < \infty.$$

Choose k_1 so small that

$$\mathbb{E}_{g_1}^{x,u} T(W^c) < u/8. \tag{6.4}$$

Let

$$V = \{(x,s) : g_1(x,s) \geq 2^{k_1}, s \in [u/2, u]\}.$$

Since the g_1-process cannot exit $D \times [0, \infty)$ through $\partial D \times [0, \infty)$, (6.4) implies

$$\mathbb{P}_{g_1}^{x,u}(T(V) > u/4) < 1/2. \tag{6.5}$$

Now let $\varepsilon = 2^{k_1-1}$. Since $0 \leq g_1 \leq 1$, the process $g_1(Y_t)$ is a martingale under $\mathbb{P}^{x,s}$, and $g_1(x,s) \geq 2^{k_1}$ for $(x,s) \in V$, we see that there is at least $2^{k_1-1}/2$ chance that Y under $\mathbb{P}^{x,s}$ will hit $M \times \{0\}$ before hitting any other part of $\partial(D \times [0, \infty))$. Thus we have for $(x,s) \in V$,

$$\mathbb{P}_{g_1}^{x,s}(Y_s \in M \times \{0\}) = \int_M (g_1(y,0)/g_1(x,s))\,\mathbb{P}^{x,s}(Y_s \in dy, T(\partial(D \times [0, \infty))) = s)$$

$$\geq \int_M \mathbb{P}^{x,s}(Y_s \in dy, T(\partial(D \times [0, \infty))) = s)$$

$$\geq 2^{k_1-1}/2.$$

This and (6.5) yield, by the strong Markov property, for all $x \in D$,

$$\mathbb{P}_{g_1}^{x,u}(Y_u \in M \times \{0\}) \geq c_{10} > 0.$$

The ratio of g and g_1 is bounded away from 0 and ∞ on the boundary of $D \times [0, \infty)$, so

$$\mathbb{P}_g^{x,u}(Y_u \in M \times \{0\}) \geq c_{11} > 0$$

for all $x \in D$. This is equivalent to the statement in the lemma. $\square$

The following theorem is a form of the boundary Harnack principle. Let $p_t^D(x,y)$ denote the transition density for Brownian motion killed on exiting D.

Theorem 6.2. *For each $u > 0$ there exists $c = c(D, u) > 0$ such that*

$$\frac{p_t^D(x,y)}{p_t^D(x,z)} \geq c\, \frac{p_s^D(v,y)}{p_s^D(v,z)}$$

for all $s, t \geq u$ and all $v, x, y, z \in D$.

Proof. First we will show that $p_u^D(x, y)$ is comparable to $\psi(x)\psi(y)$ where $\psi(x) := \mathbb{P}^x(T(D^c) > u/3)$. To simplify the notation, let us take $u = 3$.

Note that $p_1^D(\cdot, \cdot) < c$ and $p_1^D(v, z) = p_1^D(z, v)$ for all $v, z \in D$. We have

$$p_2^D(z, y) = \int_D p_1^D(z, v)p_1^D(v, y)\, dv$$

$$\le \int_D cp_1^D(v, y)\, dv = \int_D cp_1^D(y, v)\, dv$$

$$= c\psi(y).$$

It follows that

$$p_3^D(x, y) = \int_D p_1^D(x, z)p_2^D(z, y)\, dz$$

$$\le \int_D p_1^D(x, z)c\psi(y)\, dz$$

$$= c\psi(x)\psi(y).$$

It is easy to see that $p_1^D(z, v) > c_1$ for all $z, v \in M$, where M is a compact ball in D. For $z \in M$, we obtain, using Lemma 6.1,

$$p_2^D(z, y) \ge \int_M p_1^D(z, v)p_1^D(v, y)\, dv$$

$$\ge \int_M c_1 p_1^D(v, y)\, dv$$

$$= \int_M c_1 p_1^D(y, v)\, dv$$

$$= c_1 \mathbb{P}^y(X_1 \in M, T(D^c) > 1)$$

$$\ge c_1 c_2 \mathbb{P}^y(T(D^c) > 1) = c_1 c_2 \psi(y).$$

Hence, for all $x, y \in D$,

$$p_3^D(x, y) \ge \int_M p_1^D(x, z)p_2^D(z, y)\, dz$$

$$\ge \int_M p_1^D(x, z)c_1 c_2 \psi(y)\, dz$$

$$= c_1 c_2 \mathbb{P}^x(X_1 \in M, T(D^c) > 1)\psi(y)$$

$$\ge c_1 c_2^2 \mathbb{P}^x(T(D^c) > 1)\psi(y)$$

$$= c_1 c_2^2 \psi(x)\psi(y).$$

Thus, for some $c_3 > 0$ and all $x, y \in D$,

$$c_3 < p_u^D(x, y)/\psi(x)\psi(y) < c_3^{-1}.$$

This implies that

$$\frac{p_u^D(x, y)}{p_u^D(x, z)} \frac{p_u^D(v, z)}{p_u^D(v, y)} \geq \frac{c_3\psi(x)\psi(y)}{c_3^{-1}\psi(x)\psi(z)} \frac{c_3\psi(v)\psi(z)}{c_3^{-1}\psi(v)\psi(y)} = c_3^4, \qquad (6.6)$$

which proves the theorem for $s = t = u$.

In order to extend the last formula to times greater than u we use the Markov property as follows. Let $a = c_3^4 p_u^D(v, y)/p_u^D(v, z)$. Then, according to (6.6),

$$p_u^D(w, y) \geq a p_u^D(w, z)$$

for all $w, y, z \in D$. Then, for $s > u$, $x, y, z \in D$,

$$\begin{aligned}
p_s^D(x, y) &= \int_D p_{s-u}^D(x, w) p_u^D(w, y) dv \\
&\geq a \int_D p_{s-u}^D(x, w) p_u^D(w, z) dv \\
&= a p_s^D(x, z) \\
&= c_3^4 (p_u^D(v, y)/p_u^D(v, z)) p_s^D(x, z),
\end{aligned}$$

and so

$$\frac{p_s^D(x, y)}{p_s^D(x, z)} \geq c_3^4 \frac{p_u^D(v, y)}{p_u^D(v, z)}.$$

An analogous argument may be used to replace u in the right hand side with an arbitrary $t > u$ and we obtain

$$\frac{p_s^D(x, y)}{p_s^D(x, z)} \geq c_3^4 \frac{p_t^D(v, y)}{p_t^D(v, z)}$$

for all $v, x, y, z \in D$, $s, t > u$.   □

6.1 An Application of Parabolic Boundary Harnack Principle to Couplings

This section is based on [AB04]. Throughout this section, we fix a lip domain $D \subset \mathbb{R}^2$ and a sequence of lip domains D^n increasing to D, such that the defining functions f_1^n, f_2^n (cf. Eq. (5.1)) are C^2-smooth and Lipschitz with the Lipschitz constants strictly less than one. Recall that a lip domain is a Lipschitz domain, i.e., its boundary can be represented as the graph of a Lipschitz function with constant λ in some neighborhood of every boundary point (the point here is that the extreme left and right points have to be included). We will assume without loss of generality that there exist constants $\lambda < \infty$ and $r_0 > 0$, not depending on n, such that for every n and $x \in \partial D^n$, the set $\partial D^n \cap \mathcal{B}(x, r_0)$ can be represented as the graph of a (single) function with Lipschitz constant λ. Moreover, the same applies to any $x \in \partial D$ and $\partial D \cap \mathcal{B}(x, r_0)$. For each D^n, Corollary 5.6 and Propositions 5.7 and 5.8 apply.

Let $D_\varepsilon = \{x \in D : \text{dist}(x, \partial D) \geq \varepsilon\}$ and $D_\varepsilon^n = \{x \in D^n : \text{dist}(x, \partial D^n) \geq \varepsilon\}$. The hitting time of a set A by a process W will be denoted $T^W(A)$, or simply $T(A)$, if no confusion may arise.

Lemma 6.3. *Let $Q_{x,y}^n$ denote any probability measure under which (X, Y) is a strong Markov process, and X (resp., Y) is a reflected Brownian motion in D^n starting from x (resp., y). Then there are constants $\delta, \sigma > 0$, such that for any $\rho > 0$, n, $x, y \in \overline{D^n}$, $|x - y| \geq \rho$,*

$$Q_{x,y}^n(T_0 \leq \sigma\rho^2) \geq 1/2,$$

where

$$T_0 = \inf\{t > 0 : X_t, Y_t \in D_{\delta\rho}^n, X_t \in \mathcal{B}(x, \rho/16), Y_t \in \mathcal{B}(y, \rho/16)\}.$$

Let $\mathbb{P}_{x,y}^n$ be a measure under which (X, Y) is a mirror coupling of reflected Brownian motions in D^n. Let τ be the coupling time for X and Y and

$$\mathcal{D}^n = \{(x, y) \in \overline{D^n} \times \overline{D^n} : x \neq y\},$$

$$\mathcal{D}^n(\varepsilon) = \{(x, y) \in \overline{D^n} \times \overline{D^n} : |x - y| \geq \varepsilon\},$$

$$\hat{\mathcal{D}}^n(\varepsilon) = \mathcal{D}^n \setminus \mathcal{D}^n(\varepsilon).$$

Lemma 6.4. *(i) There are constants $\alpha, c, a > 0$ such that for all $\varepsilon \in (0, a)$ and all $(x, y) \in \mathcal{D}^n(\varepsilon)$,*

$$\mathbb{P}_{x,y}^n(T(\mathcal{D}^n(a)) < \tau) \geq c\varepsilon^\alpha.$$

(ii) There is a constant $c < \infty$ such that for all $\varepsilon > 0$ and $(x, y) \in \hat{\mathcal{D}}^n(\varepsilon)$,

$$\mathbb{E}_{x,y}^n T((\hat{\mathcal{D}}^n(\varepsilon))^c) \leq c\varepsilon^2.$$

(iii) For any $u > 0$ we can find $\varepsilon_0, c > 0$ such that $\mathcal{D}^1(2\varepsilon_0)$ has a non-empty interior and

$$\mathbb{P}_{x,y}^n(\tau > t, (X_t, Y_t) \in \mathcal{D}^n(\varepsilon_0)) \geq c,$$

for all n, $(x, y) \in \mathcal{D}^n(2\varepsilon_0)$, and $t \in [u/4, u]$.

See [AB04] for proofs of Lemmas 6.3 and 6.4.

Lemma 6.5 below is a version of the parabolic boundary Harnack principle for the process (X, Y). More precisely, it is a version of Lemma 6.1. We omit the proof because it is very close to that of Lemma 6.1.

Lemma 6.5. *For every $u > 0$ there exist constants $c_1, c_2 > 0$ such that for all n large and for all $x, y \in D^n$ with $x \geq y$ and $x \neq y$,*

$$\mathbb{P}^n_{x,y}(|X_u - Y_u| > c_1 \mid \tau > u) > c_2.$$

We will consider lip domains $D \subset \mathbb{R}^2$ in the rest of the section.

Recall the coordinate systems (e_1, e_2) and (e'_1, e'_2), and the partial order "$\leq$" on $\mathbb{R}^2$, defined before Proposition 5.8. For a function $u \in C^1(D)$ write

$$\partial' u = \min \left\{ \frac{\partial u}{\partial e'_1}, \frac{\partial u}{\partial e'_2} \right\}.$$

Let

$$S = \{u \in C^1(D) : \partial' u(x) \geq 0, x \in D\},$$

$$\tilde{S} = \{u \in C^1(D) : \partial' u(x) > 0, x \in D\}.$$

Lemma 6.6. *Let D be a lip domain other than a rectangle. Consider a Neumann eigenfunction φ in D corresponding to the second eigenvalue. If $\varphi \in S$ then $\varphi \in \tilde{S}$.*

Proof. We will first prove the following claim.

There exists a nonempty ball $B \subset D$ and a constant $a > 0$ (6.7)

such that $\partial' \varphi \geq a$ on B.

Assume that (6.7) does not hold. Since $\varphi \in S$ and φ is non-constant, there is $x \in D$ where $\partial \varphi(x)/\partial e'_i > 0$ for either $i = 1, 2$. Assume without loss of generality that $\partial \varphi(x)/\partial e'_1 > 0$. The inequality holds in a neighborhood of x, by continuity of $\nabla \varphi$. We have assumed that (6.7) fails, so there exists an open set where $\partial \varphi/\partial e'_2 = 0$. Since φ, as an eigenfunction, is real-analytic in D, $\partial \varphi/\partial e'_2 = 0$ holds everywhere in D. Hence $\varphi(x) = g(x_1)$, where x_1 refers to the first coordinate of $x = (x_1, x_2)$ in the coordinate system (e'_1, e'_2). The only function of the form $\varphi(x) = g(x_1)$ which satisfies the equation $\Delta \varphi = -\mu_2 \varphi$ is $\varphi(x) = \cos x_1$, upon appropriate translation and scaling of the coordinate system. As a generalized solution to the eigenfunction equation, φ satisfies for any $f \in C^2(D)$,

$$\mu_2 \int_D f\varphi \, dx + \int_D \varphi \Delta f \, dx + \int_{\partial D} \varphi \frac{\partial f}{\partial \nu} \sigma(dx) = 0,$$

where ν is the inward unit normal vector field, and σ is the surface measure on ∂D. The divergence theorem implies that

$$\mu_2 \int_D f\varphi dx = \int_D \nabla f \cdot \nabla \varphi dx.$$

We will use a test function of the form $f(x) = \theta(x_2)$. For any such function f, the right hand side of the last formula is zero, and so we have

$$\int_D \cos x_1 \, \theta(x_2) dx = 0. \tag{6.8}$$

The eigenfunction φ must vanish at some points of D. It follows from the periodicity of $\varphi(x) = \cos x_1$ that we can assume without loss of generality that φ vanishes on $\{x \in D : x_1 = \pi/2\}$. By Courant's Nodal Line Theorem, φ cannot vanish anywhere else in D. Hence, $D \subset \{x : -\pi/2 < x_1 < 3\pi/2\}$. Let x^* and y^* be the left and right vertices of D. Let α_1 [resp., α_2] denote the line through x^* [resp., y^*], parallel to the e_1'-axis. The fact that D is a lip domain which is not a rectangle implies that the projection onto the e_1'-axis of $\alpha_1 \cap \partial D$ is not equal to that of $\alpha_2 \cap \partial D$. Hence, for either $i = 1$ or 2, $\alpha_i \cap \partial D$ is not symmetric about the line through the point $(\pi/2, 0)$ parallel to e_2'. Assume without loss of generality that there is no symmetry for $i = 1$. Then either $\int_{\alpha \cap D} \varphi(x_1, x_2) dx_1 > 0$ for all lines α which intersect D and are sufficiently close to α_1, or the integral is strictly negative for all such α. Now let θ be such that $f(x) = \theta(x_2)$ is zero off an ε-neighborhood of α_1, it is equal to 1 in a neighborhood of α_1, and nonnegative in D. Then, if ε is small enough, $\int_D f\varphi dx \neq 0$. This contradicts (6.8), and therefore (6.7) holds.

We will need the following result [BC02, Theorem 2.4] on "synchronous couplings." Suppose $x, y \in D$. Then there is a probability space $(\Omega', \mathcal{F}', \mathbb{P}'_{x,y})$, and a process (W, X, Y) such that X [respectively, Y] is a reflected Brownian motion in $\overline{D}$ starting from x [y], and W is a Brownian motion on the filtration generated by (W, X, Y), such that (X, Y) admits the following Skorohod representation:

$$X_t = x + W_t + \int_0^t \mathbf{n}(X_s) d|L|_s, \quad Y_t = y + W_t + \int_0^t \mathbf{n}(Y_s) d|M|_s,$$

where $|L|$ and $|M|$ are boundary local times for X and Y, respectively, and where $\mathbf{n}$ is the inward normal vector field on ∂D. Theorem 2.4 of [BC02] shows that the synchronous coupling (X, Y) in D may be obtained as a limit of synchronous couplings in a sequence of polygonal lip domains D_k approximating D. For polygonal lip domains, the following "order preserving property" has been proved in [BB99], and so it holds for the coupling (X, Y) in D; if $x \leq y$ then $X_t \leq Y_t$ for all $t \geq 0$, $\mathbb{P}'_{x,y}$-a.s. We have for any $t > 0$,

$$e^{-\mu_2 t} \varphi(x) = \mathbb{E}'_{x,y} \, \varphi(X_t), \quad e^{-\mu_2 t} \varphi(y) = \mathbb{E}'_{x,y} \, \varphi(Y_t).$$

Let $B = B_\rho(x_0)$ be a ball satisfying claim (6.7) proved at the beginning of the proof, and let $\tilde{B} = B_{\rho/2}(x_0)$. Let $x \in D$, $y = x + re'_1$ and $T = \inf\{t : \text{dist}(X_t, \partial D) < \text{dist}(x, \partial D)/2\}$. Note that $\varphi(Y_1) - \varphi(X_1) \geq 0$ with probability 1, because $X_1 \leq Y_1$ and $\varphi \in S$. We have $Y_1 - X_1 = Y_0 - X_0$ if $T > 1$. Hence for all $r \in (0, \rho/2)$ such that $y \in D$,

$$r^{-1}(\varphi(y) - \varphi(x)) \geq r^{-1}e^{\mu_2} \mathbb{E}'_{x,y}\left[(\varphi(Y_1) - \varphi(X_1))1_{\{X_1 \in \tilde{B}, T > 1\}}\right]$$

$$\geq (e^{\mu_2} a/2) \mathbb{P}'_{x,y}(X_1 \in \tilde{B}, T > 1),$$

where $a > 0$ is the constant in (6.7). Since the right hand side is positive and independent of r, it follows that $\partial\varphi(x)/\partial e'_1 > 0$. A similar argument applies to $\partial\varphi/\partial e'_2$, and we conclude that $\partial'\varphi > 0$ on D. $\square$

Recall that $D_\varepsilon = \{x \in D : \text{dist}(x, \partial D) \geq \varepsilon\}$.

Lemma 6.7. *Suppose D is a lip domain other than a rectangle. For any $\varepsilon_1 > 0$ such that the interior of D_{ε_1} is non-empty, and any $\delta, b > 0$, there exists $\varepsilon_2 > 0$ with the following property. If φ is a Neumann eigenfunction corresponding to μ_2 with*

$$\begin{cases} \partial'\varphi \geq \delta & \text{on } D_{\varepsilon_1}, \\ \partial'\varphi \geq 0 & \text{on } D_{\varepsilon_2}, \\ |\varphi| \leq b & \text{on } D, \end{cases}$$

then

$$\partial'\varphi \geq 0 \quad \text{on } D.$$

Proof. Let $\mathbb{E}_x$ be the expectation corresponding to the distribution of reflected Brownian motion in D, starting from x. Then for any $t \geq 0$, and $x, y \in \overline{D}$,

$$e^{-\mu_2 t}(\varphi(x) - \varphi(y)) = \mathbb{E}_x \varphi(X_t) - \mathbb{E}_y \varphi(Y_t).$$

Hence, it will suffice to show that for all $x, y \in D$, $x \geq y$,

$$\mathbb{E}_x \varphi(X_2) - \mathbb{E}_y \varphi(Y_2) \geq 0.$$

For a fixed lip domain $D \subset \mathbb{R}^2$, consider a sequence of lip domains D^n increasing to D, such that the defining functions f_1^n, f_2^n (cf. Eq. (5.1)) are C^2-smooth and Lipschitz with the Lipschitz constants strictly less than one. We will denote by $\mathbb{P}^n_{x,y}$ a measure under which (X, Y) is a pair of reflected Brownian motions in D^n, starting from (x, y), which is mirror coupled on the time interval $[0, 1)$ and synchronously coupled on $[1, 2]$.

If $x \in D$ then the distributions of X under $\mathbb{P}^n_{x,y}$ converge weakly to the distribution of the reflected Brownian motion in D in the uniform topology on $C([0,2], \mathbb{R}^2)$, as $n \to \infty$, by [BC98, Theorem 2]. Since a similar remark applies to Y, it is enough to show that there exists $\varepsilon_2 > 0$ such that for all large n,

$$\mathbb{E}^n_{x,y}(\varphi(X_2) - \varphi(Y_2)) \geq 0.$$

Denote the coupling time by τ. Since $X_2 = Y_2$ on $\{\tau \leq 1\}$, it suffices to show that

$$\mathbb{E}^n_{x,y}[\varphi(X_2) - \varphi(Y_2) \mid \tau > 1] \geq 0. \tag{6.9}$$

By the uniform continuity of transition probabilities in D^n's proved in [BC02, Theorem 2.1],

$$\mathbb{P}^n_{x,y}(X_1 \in C) \leq c|C|, \tag{6.10}$$

for any Borel subset C of $\overline{D}$, where $|C|$ denotes the Lebesgue measure of C, and c does not depend on x, C and n. Let c_1 and c_2 be the constants from the statement of Lemma 6.5. We can assume that $\varepsilon_2 < \varepsilon_1$. By Proposition 5.8, $X_2 \geq Y_2$ a.s. Recall that $\partial'\varphi \geq 0$ on D_{ε_2}. This implies that $\varphi(X_2) - \varphi(Y_2) \geq 0$ on $\{X_2, Y_2 \in D_{\varepsilon_2}\}$. By (6.10),

$$\mathbb{E}^n_{x,y}(\varphi(X_2) - \varphi(Y_2)|\tau > 1) \geq \mathbb{E}^n_{x,y}[(\varphi(X_2) - \varphi(Y_2))1_{\{X_2,Y_2 \in D_{\varepsilon_2}\}}|\tau > 1]$$
$$-2b[\mathbb{P}_{x,y}(X_2 \notin D_{\varepsilon_2}|\tau > 1) + \mathbb{P}_{x,y}(Y_2 \notin D_{\varepsilon_2}|\tau > 1)]$$
$$\geq \mathbb{E}^n_{x,y}[(\varphi(X_2) - \varphi(Y_2))1_{\{X_2,Y_2 \in D_{\varepsilon_1}, |X_2-Y_2|>c_1/2\}}|\tau > 1]$$
$$-c\varepsilon_2.$$

Since $\partial'\varphi \geq \delta$ on D_{ε_1}, we have $\varphi(X_2) - \varphi(Y_2) \geq \delta c_1/2$ on $\{X_2, Y_2 \in D_{\varepsilon_1}, |X_2 - Y_2| > c_1/2\}$, so

$$\mathbb{E}^n_{x,y}(\varphi(X_2)-\varphi(Y_2)|\tau > 1) \geq c\,\mathbb{P}^n_{x,y}(X_2, Y_2 \in D_{\varepsilon_1}, |X_2-Y_2| > c_1/2 \mid \tau > 1)-c\varepsilon_2.$$

It is clear that Lemma 6.5 remains valid if we replace c_1 in that lemma with any smaller constant. Assume that $c_1 > 0$ is so small that there exists $z \in D$ such that $B_{2c_1}(z) \subset D_{\varepsilon_1}$. An easy argument based on Lemma 6.3 and the strong Markov property shows that if $|X_1 - Y_1| > c_1$ then with some probability $p > 0$, independent of n, we have $X_2, Y_2 \in D_{\varepsilon_1}$ and $|X_2 - Y_2| > c_1/2$. Hence,

$$\mathbb{E}^n_{x,y}(\varphi(X_2) - \varphi(Y_2)|\tau > 1) \geq c\,\mathbb{P}^n_{x,y}[|X_1 - Y_1| > c_1|\tau > 1] - c\varepsilon_2$$
$$\geq cc_2 - c\varepsilon_2,$$

where the last inequality follows from Lemma 6.5. Note that c_2 and the other constants in the last formula do not depend on x and y. Taking $\varepsilon_2 > 0$ small, (6.9) follows, and Lemma 6.7 is proved. $\square$

Theorem 6.8. *(i) The second eigenvalue μ_2 is simple in all lip domains except squares.*

(ii) ("Hot spots conjecture") For every lip domain, every Neumann eigenfunction corresponding to μ_2 attains its maximum and minimum at boundary points only.

Proof. One can easily verify that both parts of the theorem hold in rectangles so we will assume that D is not a rectangle.

(i) The argument is similar to that in the proof of [Ata01, Theorem 4.1]. Arguing by contradiction, assume that μ_2 is multiple.

Recall from the proof of Lemma 6.6 that there exists a synchronous coupling of reflecting Brownian motions in D with the property that $X_t \leq Y_t$ for all $t \geq 0$, if $X_0 \leq Y_0$. This and the argument given in the proof of [BB99, Theorem 3.3] show that at least one of the eigenfunctions corresponding to μ_2 belongs to S. Fix one of these eigenfunctions and call it φ.

We have assumed that μ_2 is multiple so there exists an eigenfunction $\varphi^\perp$ which is orthogonal to φ. Since $\varphi^\perp$ and $-\varphi^\perp$ cannot both be in S, we can assume without loss of generality that $\varphi^\perp \notin S$. Let

$$\varphi^a = (1-a)\varphi + a\varphi^\perp, \quad a \in [0,1],$$

and

$$a^* = \inf\{a \in [0,1] : \varphi^a \notin S\}.$$

We claim that $a^* < 1$. Since $\partial'\varphi^a \geq 0$ on D for $a < a^*$, and $\partial'\varphi^a \to \partial'\varphi^{a^*}$ pointwise in D when $a \uparrow a^*$, we have that $\varphi^{a^*} \in S$. Hence a^* cannot be equal to 1. Therefore there is a sequence a_k with

$$a_k \downarrow a^*, \quad a_k \in (a^*, 1), \quad \varphi^{a_k} \notin S. \tag{6.11}$$

For $a \in (a^*, 1)$ let

$$\varepsilon(a) = \sup\{\text{dist}(x, \partial D) : \partial'\varphi^a(x) < 0\}.$$

Note that as $a \downarrow a^*$, $\partial'(\varphi^a) \to \partial'(\varphi^{a^*})$ uniformly on compact subsets of D. Since $\varphi^{a^*} \in S$, Lemma 6.6 implies that $\varphi^{a^*} \in \tilde{S}$. Therefore for any compact set C, we have $\partial'(\varphi^a) > 0$ on C for all a sufficiently close to a^*. Hence

$$\lim_{a\downarrow a^*} \varepsilon(a) = 0. \tag{6.12}$$

Let $\varepsilon_1 > 0$ be as in Lemma 6.7. Using again the facts that the convergence $\partial' \varphi^a \to \partial' \varphi^{a^*}$ is uniform on compacts, and that $\varphi^{a^*} \in \tilde{S}$, we have that there are constants $\delta > 0$ and $a_1 \in (a^*, 1)$ such that $\partial' \varphi^a \geq \delta$ on D_{ε_1} for all $a \in (a^*, a_1)$. Since by definition, $\partial' \varphi^a \geq 0$ on $D_{\varepsilon(a)}$ for all $a \in (a^*, a_1)$, Lemma 6.7 and (6.12) show that $\varphi^a \in S$ for all $a \in (a^*, a_2)$, where $a_2 \in (a^*, a_1)$. This gives a contradiction with (6.11). As a result, the eigenvalue is simple, and Theorem 6.8 (i) follows.

(ii) We have shown in the first part of the proof that there is only one eigenfunction corresponding to μ_2 and that it belongs to S. This immediately implies part (ii) of the theorem.

$\square$

Chapter 7
Scaling Coupling

This chapter is based on a paper by Pascu [Pas02]. In this chapter, we denote the unit disk in $\mathbb{R}^2$ by $U = \{z : |z| < 1\}$.

A curve $\Gamma \subset \mathbb{R}^2$ is said to be of class $C^{1,\alpha}$ $(0 < \alpha < 1)$ if it has a parametrization $w(t)$ that is continuously differentiable, $w' \neq 0$ and w' is Hölder of order α, that is, for some $M > 0$ and for all t, t' we have:

$$\left| w'(t) - w'(t') \right| \leq M \left| t - t' \right|^\alpha .$$

A domain $D \subset \mathbb{R}^2$ is said to be a $C^{1,\alpha}$ domain $(0 < \alpha < 1)$ if its boundary is a Jordan curve Γ of class $C^{1,\alpha}$. It is known [Pom92, pp. 48–49] that if $f : U \to D$ is a conformal map of the unit disk onto the $C^{1,\alpha}$ domain D $(0 < \alpha < 1)$, then f and f' have continuous extensions to $\overline{U}$, f' is Hölder of order α on $\overline{U}$ and $f' \neq 0$ on $\overline{U}$. An analytic function f is called convex in U if it maps U conformally onto a convex domain.

The inward unit normal vector field on ∂D will be denoted $\mathbf{n}_D$. Recall that reflected Brownian motion in D can be defined as a solution of the stochastic differential equation:

$$X_t = X_0 + B_t + \frac{1}{2} \int_0^t \mathbf{n}_D(X_s) dL_s, \qquad t \geq 0, \tag{7.1}$$

where B_t is a two-dimensional Brownian motion started at 0 and L_t is a continuous nondecreasing process which increases only when $X_t \in \partial D$, and almost surely $X_0 = x_0$ and $X_t \in \overline{D}$ for all $t \geq 0$.

K. Burdzy, *Brownian Motion and its Applications to Mathematical Analysis*,
Lecture Notes in Mathematics 2106, DOI 10.1007/978-3-319-04394-4_7,
© Springer International Publishing Switzerland 2014

7.1 Scaling Coupling of Reflected Brownian Motions in the Unit Disk

We will first construct the scaling coupling in the case of the unit disk and then extend it to bounded $C^{1,\alpha}$ domains by means of conformal maps.

Theorem 7.1. *Let Z_t be a reflected Brownian motion in U starting at $z_0 \in \overline{U} - \{0\}$ and consider any $a \in [|z_0|, 1]$. The process $\tilde{Z}_t$ defined by:*

$$\tilde{Z}_t = \frac{1}{M_{\gamma_t}} Z_{\gamma_t}, \qquad t \geq 0, \tag{7.2}$$

where:

$$M_t = a \vee \sup_{s \leq t} |Z_s|,$$

$$C_t = \int_0^t \frac{1}{M_s^2} ds, \tag{7.3}$$

$$\gamma_t = \inf\{s > 0 : C_s \geq t\}, \tag{7.4}$$

is reflected Brownian motion in U starting at $\tilde{z}_0 = z_0/a$.

Proof. We apply Itô's formula to the semimartingale Z_t and the nondecreasing process M_t with $f(x, y) = x/y$. If

$$Z_t = Z_0 + B_t + \frac{1}{2} \int_0^t \mathbf{n}_U(Z_s) dL_s$$

is the semimartingale representation of Z_t given by (7.1), we have:

$$\frac{Z_t}{M_t} = \frac{Z_0}{M_0} + \int_0^t \frac{1}{M_s} dZ_s - \int_0^t \frac{1}{M_s^2} Z_s dM_s \tag{7.5}$$

$$= \frac{Z_0}{M_0} + \int_0^t \frac{1}{M_s} dB_s + \frac{1}{2} \int_0^t \frac{1}{M_s} \mathbf{n}_U(Z_s) dL_s - \int_0^t \frac{1}{M_s^2} Z_s dM_s.$$

If $\tau = \inf\{s : |Z_s| = 1\}$, note that L_s is constant on $[0, \tau]$ and $M_s \equiv 1$ on $[\tau, \infty)$; further, when M_s is increasing, Z_s/M_s is on ∂U, and therefore $-Z_s/M_s = \mathbf{n}_U(Z_s/M_s)$. The difference of the last two integrals in the last equality above can thus be written:

$$\frac{1}{2}\int_{t\wedge\tau}^{t}\frac{1}{M_s}\mathbf{n}_U(Z_s)dL_s - \int_{0}^{t\wedge\tau}\frac{1}{M_s^2}Z_s dM_s$$

$$=\frac{1}{2}\int_{t\wedge\tau}^{t}\mathbf{n}_U(Z_s/M_s)dL_s + \frac{1}{2}\int_{0}^{t\wedge\tau}\mathbf{n}_U(Z_s/M_s)d\log M_s^2$$

$$=\frac{1}{2}\int_{0}^{t}\mathbf{n}_U(Z_s/M_s)dL_s + \frac{1}{2}\int_{0}^{t}\mathbf{n}_U(Z_s/M_s)d\log M_s^2$$

$$=\frac{1}{2}\int_{0}^{t}\mathbf{n}_U(Z_s/M_s)d\tilde{L}_s,$$

where $\tilde{L}_s = L_s + \log M_s^2$ is readily seen to be a nondecreasing process which increases only if either L_s or M_s do; this happens only when Z_s/M_s is on the boundary of U.

Note that since $0 < |z_0| \le M_t \le 1$ for all $t \ge 0$, C_t defined by (7.3) is strictly increasing and $C_t \to \infty$ a.s. It follows that γ_t defined by (7.4) is continuous and increasing, and substituting in (7.5) we obtain:

$$\tilde{Z}_t = \frac{Z_{\gamma_t}}{M_{\gamma_t}} = \frac{Z_0}{M_0} + \int_{0}^{\gamma_t}\frac{1}{M_s}dB_s + \frac{1}{2}\int_{0}^{\gamma_t}\mathbf{n}_U(Z_s/M_s)d\tilde{L}_s.$$

Setting $\tilde{B}_t = \int_{0}^{\gamma_t}\frac{1}{M_s}dB_s$, we obtain:

$$\langle\tilde{B}^i, \tilde{B}^j\rangle_t = \int_{0}^{\gamma_t}\frac{1}{M_s^2}d\langle\tilde{B}^i, \tilde{B}^j\rangle_s = \delta_{ij}\int_{0}^{\gamma_t}\frac{1}{M_s^2}ds = \delta_{ij}C_{\gamma_t} = \delta_{ij}t,$$

$i, j \in \{1, 2\}$, and thus $\tilde{B}_t$ is a two-dimensional Brownian motion starting at 0.

We have thus shown:

$$\tilde{Z}_t = \tilde{Z}_0 + \tilde{B}_t + \frac{1}{2}\int_{0}^{\gamma_t}\mathbf{n}_D(Z_s/M_s)d\tilde{L}_s$$

$$= \tilde{Z}_0 + \tilde{B}_t + \frac{1}{2}\int_{0}^{t}\mathbf{n}_D(Z_{\gamma_u}/M_{\gamma_u})d\tilde{L}_{\gamma_u}$$

$$= \tilde{Z}_0 + \tilde{B}_t + \frac{1}{2}\int_{0}^{t}\mathbf{n}_D(\tilde{Z}_u)d\overline{L}_u,$$

where $\overline{L}_u = \tilde{L}_{\gamma_u}$ is a nondecreasing process which increases only when $\tilde{Z}_u = Z_{\gamma_u}/M_{\gamma_u}$ is at the boundary of U. This proves the theorem.   □

Definition 7.2. We call the pair $(Z_t, \tilde{Z}_t)$ constructed above a *scaling coupling* of reflected Brownian motions in U, starting at $z_0 \in \overline{U} - \{0\}$ and at $\tilde{z}_0 = \frac{1}{a}z_0 \in \overline{U}$.

Remark 7.3. It was proved in [Pas02] that the conformal image of a reflected Brownian motion in the unit disk is a time change of a reflected Brownian motion in the image domain, provided the target domain is $C^{1,\alpha}$. In fact, the claim is true even if the target domain D is a Jordan domain, that is, $\overline{D}$ is homeomorphic to the closed unit disk. The following outline of the proof of the more general claim is due to Zhenqing Chen; it appeared in [BK98, Sect. 2].

Suppose that Z_t is reflected Brownian motion in U starting at $z_0 \in \overline{U}$ and f is a conformal map of U onto a Jordan domain D. We will argue that $f(Z_t)$ is a reflected Brownian motion in D, up to a time-change. First, the reflected Brownian motion on a simply connected Jordan domain D can be characterized as the continuous Markov process associated with $(H^1(D), \mathcal{E})$ in $L^2(D, dz)$, which is a regular Dirichlet form on $\overline{D}$, where $\mathcal{E}(f, g) = 1/2 \int_D \nabla f \nabla g \, dz$ (see Remark 1 to Theorem 2.3 of [Che92b]). We note that if $(H^1(D), \mathcal{E})$ is regular on $\overline{D}$ then the associated process is the unique continuous strong Markov process in $\overline{D}$ that is reversible with respect to the Lebesgue measure, behaves like planar Brownian motion in D and spends zero time on the boundary.

The Dirichlet form for the process $f(Z_t)$ under the reference measure $|f'(z)|^2 dz$ is $(H^1(f^{-1}(D)), \mathcal{E})$. Therefore $f(Z_t)$ is a time change of the reflected Brownian motion on D. The last assertion follows from a Dirichlet form characterization of time-changed processes due to Silverstein and Fitzsimmons (see [Sil74, Theorems 8.2 and 8.5]; that proof seems to contain a gap; see [Fit89] for a correct proof; see also [FŌT94, Theorem 6.2.1]).

7.2 Scaling Coupling of Reflected Brownian Motions in $C^{1,\alpha}$ Domains

Definition 7.4. (i) We define a hyperbolic line in U as being a line segment or an arc of a circle contained in $\overline{U}$ which meets orthogonally the boundary of U. We denote by $\mathcal{H}_U$ the family of all hyperbolic lines in U. If z_1, z_2 are two distinct points on a hyperbolic line $l \in \mathcal{H}_U$, we define the hyperbolic segment with endpoints z_1 and z_2 (denoted by $[z_1, z_2]$) as the part of l between (and including) z_1 and z_2.

(ii) For a $C^{1,\alpha}$ $(0 < \alpha < 1)$ domain D, we define a hyperbolic line/segment in D as the conformal image of a hyperbolic line/segment in U. We denote by $\mathcal{H}_D$ the family of all hyperbolic lines in D.

Simple geometric considerations show the following:

Proposition 7.5. *Let D be a $C^{1,\alpha}$ $(0 < \alpha < 1)$ domain.*

(i) Given two distinct points $\overline{D}$, there exists a unique hyperbolic line in D passing through them.

(ii) For an arbitrarily chosen diameter d of U, we have

$$\mathcal{H}_U = \{\psi(d) : \psi \in \mathcal{A}\},$$

where $\mathcal{A}$ is the family of all automorphisms of U. If f is an arbitrarily chosen conformal map of U onto D, then

$$\mathcal{H}_D = \{f \circ \psi(d) : \psi \in \mathcal{A}\}.$$

We will denote by $z_1 z_2$ the unique hyperbolic line passing through z_1 and z_2.

We give now the construction of scaling coupling for general $C^{1,\alpha}$ domains. Let D be a $C^{1,\alpha}$ domain $(0 < \alpha < 1)$ and let $w_0, \tilde{w}_0 \in \overline{D}$ be distinct and not both on ∂D. By Proposition 7.5, there is a unique hyperbolic line $w_0 \tilde{w}_0$ in D, passing through w_0 and $\tilde{w}_0$. Consider a point w_1 on $w_0 \tilde{w}_0 - [w_0, \tilde{w}_0]$, $w_1 \notin \partial D$. Let $f : U \to D$ be the unique conformal map of U onto D (given by the Riemann mapping theorem) with $f(0) = w_1$ and $\arg f'(0) = 0$. Let $z_0 = f^{-1}(w_0)$ and $\tilde{z}_0 = f^{-1}(\tilde{w}_0)$. Note that by definition, $f^{-1}(w_0 \tilde{w}_0)$ is a hyperbolic line in U, and since $0 = f^{-1}(w_1) \in f^{-1}(w_0 \tilde{w}_0)$, it follows that $f^{-1}(w_0 \tilde{w}_0)$ is in fact a diameter of U. Note that by the choice of w_1, we have $0 \notin [z_0, \tilde{z}_0]$, and therefore $|z_0| \neq |\tilde{z}_0|$. Without loss of generality we can assume that $|z_0| < |\tilde{z}_0|$.

Let Z_t be a reflecting Brownian motion in U starting at z_0. Define processes W_t, $\tilde{W}_t$ by:

$$W_t = f(Z_{\alpha_t}), \qquad t \geq 0, \tag{7.6}$$

$$\tilde{W}_t = f\left(\frac{1}{M_{\beta_t}} Z_{\beta_t}\right), \qquad t \geq 0, \tag{7.7}$$

where $M_t = |z_0/\tilde{z}_0| \vee \sup_{s \leq t} |Z_s|, t \geq 0$ and

$$A_t = \int_0^t |f'(Z_s)|^2 \, ds, \qquad \alpha_t = \inf\{s : A_s \geq t\}, \qquad t \geq 0, \tag{7.8}$$

$$B_t = \int\limits_0^t \frac{1}{M_s^2} \left| f'\left(Z_s/M_s\right) \right|^2 ds, \qquad \beta_t = \inf\{s : B_s \geq t\}, \qquad t \geq 0. \quad (7.9)$$

Theorem 7.6. W_t, $\tilde{W}_t$ defined by (7.6)–(7.9) are reflected Brownian motions in D, starting at w_0, respectively $\tilde{w}_0$.

Proof. Remark 7.3 implies that W_t is reflected Brownian motion in D.

To prove that $\tilde{W}_t$ is reflected Brownian motion, note that by Lemma 7.1, Z_t/M_t is a time change γ_t (given by (7.4)) of a reflected Brownian motion in $\overline{U}$, starting at $\tilde{z}_0$.

By Remark 7.3, $f(Z_{\gamma_t}/M_{\gamma_t})$ is a time change $\tilde{\alpha}_t$ of a reflected Brownian motion in D, starting at $f(\tilde{z}_0) = \tilde{w}_0$, where

$$\tilde{\alpha}_t = \inf\{s : \tilde{A}_s \geq t\} \text{ and } \tilde{A}_t = \int\limits_0^t \left| f'\left(\frac{Z_{\gamma_s}}{M_{\gamma_s}}\right) \right|^2 ds, \qquad t \geq 0. \quad (7.10)$$

In order to prove the claim it suffices to show that the combined effect of the two time changes γ_t and $\tilde{\alpha}_t$ is the time change β_t given by (7.9), that is $\gamma_{\tilde{\alpha}_t} = \beta_t$ for all $t \geq 0$.

C_u given by (7.4) is a bijection on $[0, \infty)$, with inverse $C^{-1} = \gamma$. With the substitution $s = C_u$ in the definition of $\tilde{A}_t$, we obtain:

$$\tilde{A}_{C_t} = \int\limits_0^{C_t} \left| f'\left(\frac{Z_{\gamma_s}}{M_{\gamma_s}}\right) \right|^2 ds = \int\limits_0^t \left| f'\left(\frac{Z_u}{M_u}\right) \right|^2 \frac{dC_u}{du} du$$

$$= \int\limits_0^t \frac{1}{M_u^2} \left| f'\left(\frac{Z_u}{M_u}\right) \right|^2 du = B_t,$$

for all $t \geq 0$. This shows that $\tilde{A}_{C_t} = B_t$ for all $t \geq 0$, or equivalently, by taking inverses, we have $\gamma_{\tilde{\alpha}_t} = \beta_t$ for all $t \geq 0$, as needed. □

Definition 7.7. For a $C^{1,\alpha}$ domain D ($0 < \alpha < 1$) and any distinct points $w_0, \tilde{w}_0 \in \overline{D}$ (not both on ∂D), $w_1 \in w_0\tilde{w}_0 - [w_0, \tilde{w}_0]$ (not on ∂D), the pair $(W_t, \tilde{W}_t)$ defined by (7.6)–(7.9) will be called a *scaling coupling* of reflected Brownian motions in D starting at $w_0 \in \overline{D}$ and $\tilde{w}_0 \in \overline{D}$.

Remark 7.8. The above construction of scaling coupling for a $C^{1,\alpha}$ domain with starting points $(w_0, \tilde{w}_0)$ relies on the choice of a conformal map from the unit disk U onto D. The choice is uniquely determined by the values of $f(0)$ and $\arg f'(0)$. The choice of just $w_1 = f(0)$ determines uniquely the conformal map, up to a rotation of the unit disk. By the angular symmetry of the construction of scaling coupling in the case of the unit disk, it follows that the construction is invariant under rotations of the unit disk, and therefore the construction does not depend of the choice of arg $f'(0)$ (we chose arg $f'(0) = 0$ for simplicity).

It follows that given any two distinct points in $\overline{D}$ (not both on the boundary of D), the scaling coupling with these starting points is uniquely determined once a choice for $f(0)$ (lying on the hyperbolic line passing through them, and not separating them) has been made. We will therefore refer to $w_1 = f(0)$ as the parameter of the scaling coupling.

In order to derive the main property of the scaling coupling in the case of convex domains, we need the following characterization of convexity:

Proposition 7.9. *Let* $f : U \rightarrow D$ *be a conformal map of* U *onto the simply connected domain* D. *The following are equivalent:*

$$D \text{ is convex;} \tag{7.11}$$

$$|rf'(re^{i\theta})| \text{ is an increasing function of } r \in [0, 1), \text{ for all } 0 \leq \theta < 2\pi. \tag{7.12}$$

Proof. Since $f'(0) \neq 0$, without loss of generality we can assume that $f(0) = f'(0) - 1 = 0$.

Note that the domain D is convex iff the function f is convex, which (under the condition $f(0) = f'(0) - 1 = 0$) is equivalent (see [Dur83], p. 42) to:

$$\text{Re}\left(1 + z\frac{f''(z)}{f'(z)}\right) > 0, \quad z \in U.$$

In polar coordinates, $z = re^{i\theta}$, this is equivalent to

$$\text{Re}\left(\frac{1}{r} + e^{i\theta}\frac{f''(re^{i\theta})}{f'(re^{i\theta})}\right) > 0, \quad 0 \leq \theta < 2\pi, \quad 0 < r < 1.$$

Note that since

$$\text{Re}\left(\frac{1}{r} + e^{i\theta}\frac{f''(re^{i\theta})}{f'(re^{i\theta})}\right) = \frac{1}{r} + \text{Re}\left(\frac{\partial}{\partial r}\log f'(re^{i\theta})\right) = \frac{1}{r} + \frac{\partial}{\partial r}\ln|f'(re^{i\theta})|$$

$$= \frac{\partial}{\partial r}\ln|rf'(re^{i\theta})|,$$

the previous statement is equivalent to (7.12), as needed.   □

The crucial property of the scaling coupling is given by the following:

Proposition 7.10. *With the notation of Theorem 7.6, there exist almost surely finite stopping times* τ_1 *and* τ_2 *such that for all* $t \geq 0$ *we have*

$$\alpha_{t+\tau_1} = \beta_{t+\tau_2}.$$

Moreover, if the domain D is convex, with probability one we have $\beta_t \le \alpha_t$, for all $t \ge 0$.

Proof. Set $\tau = \inf\{s : |Z_s| = 1\}$, $\tau_1 = A_\tau$ and $\tau_2 = B_\tau$. Obviously τ is an a.s. finite stopping time and we have $M_s \equiv 1$ for all $s \ge \tau$.

It follows that τ_1 and τ_2 are also a.s. finite stopping times and that for all $t \ge \tau$ we have:

$$A_t - \tau_1 = \int_\tau^t |f'(Z_s)|^2 \, ds = B_t - \tau_2.$$

Since $\alpha_t = A_t^{-1}$, $\beta_t = B_t^{-1}$, this implies the first part of the proposition.
For the second part, note that since D is convex, Proposition 7.9 shows that:

$$|Z_s f'(Z_s)| \le \left| \frac{Z_s}{M_s} f'\left(\frac{Z_s}{M_s}\right) \right|,$$

hence we obtain:

$$A_t = \int_0^t |f'(Z_s)|^2 \, ds \le \int_0^t \frac{1}{M_s^2} \left| f'\left(\frac{Z_s}{M_s}\right) \right|^2 \, ds = B_t,$$

and therefore we have $\alpha_t \ge \beta_t$, for all $t \ge 0$. □

7.3 An Application of Scaling Couplings to the Hot Spots Problem

Let $D \subset \mathbb{R}^2$ be a convex $C^{1,\alpha}$ domain ($0 < \alpha < 1$) having a line of symmetry. Without loss of generality we will assume that D is symmetric with respect to the horizontal axis.

Set $D^+ = D \cap \{(x, y) : y > 0\}$, $\Gamma^+ = \partial D^+ \cap \partial D$, $\Gamma_0 = \overline{D} \cap \{(x, y) : y = 0\}$. For $w_0 \in \overline{D^+}$, denote by $\tau^{w_0} = \inf\{s : \operatorname{Im} W_s = 0\}$, where W_s is a reflected Brownian motion in D starting at w_0.

Remark 7.11. Let f_1 be a conformal map of U^+ onto D^+, such that the parts of the boundaries of U^+ and D^+ lying on the horizontal axis correspond to each other. By the symmetry principle, f_1 extends to a conformal map of U onto D. Since $\Gamma_0 = f_1([-1, 1])$, it follows that Γ_0 is a hyperbolic line in $\overline{D}$.

Lemma 7.12. *Given $\tilde{w}_0 \in \overline{D}$, $w_1 \in D \cap \Gamma_0$ and $w_0 \in [w_1, \tilde{w}_0]$, there exist reflected Brownian motions W_t, $\tilde{W}_t$ in D, starting at w_0, respectively $\tilde{w}_0$, such that if τ^{w_0},*

$\tau^{\tilde{w}_0}$ are the hitting times to the horizontal axis of W_t, respectively $\tilde{W}_t$, then with probability one we have:

$$\tau^{w_0} \leq \tau^{\tilde{w}_0}.$$

Proof. The proof is trivial if $w_0 = \tilde{w}_0$ or $w_0 = w_1$, so we can assume that $w_0, \tilde{w}_0$ and w_1 are distinct.

Let $(W_t, \tilde{W}_t)$ be a scaling coupling of reflected Brownian motions in D starting at $(w_0, \tilde{w}_0)$, with parameter w_1, that is a scaling coupling obtained as the image under a conformal map $f : U \rightarrow D$ with $f(0) = w_1$ of the scaling coupling $(Z_t, \tilde{Z}_t)$ in U.

Remark 7.11 above shows that Γ_0 is a hyperbolic line in U, hence $f^{-1}(\Gamma_0)$ is a hyperbolic line in U. Since $0 = f^{-1}(w_1) \in f^{-1}(\Gamma_0)$, it follows that $f^{-1}(\Gamma_0)$ is in fact a diameter of U. By Remark 7.8, without loss of generality we can assume that $f^{-1}(\Gamma_0) = [-1, 1]$.

If $s > 0$ is such that $\mathrm{Im}\, \tilde{W}_s = 0$, then, by construction, we have $\mathrm{Im}\, f(Z_{\beta_s}/M_{\beta_s}) = 0$. Because under f the parts of the boundaries of U^+ and D^+ lying on the horizontal axis correspond to each other, $\mathrm{Im}(Z_{\beta_s}/M_{\beta_s}) = 0$. Since M_{β_s} is real, it follows that $\mathrm{Im}\, Z_{\beta_s} = 0$.

The set D is convex so, by Proposition 7.10, it follows that $\beta_s \leq \alpha_s$. Since α_s is increasing (and a bijection) on $[0, \infty)$, there exists $s' \leq s$ such that $\alpha_{s'} = \beta_s$.

It follows that $\mathrm{Im}\, Z_{\alpha_{s'}} = \mathrm{Im}\, Z_{\beta_s} = 0$, hence $W_{s'} = f(Z_{\alpha_{s'}})$ is on the horizontal axis.

We have shown that if $\mathrm{Im}\, \tilde{W}_s = 0$, then there exists $s' \leq s$ such that $\mathrm{Im}\, W_{s'} = 0$, which implies that $\tau^{w_0} \leq \tau^{\tilde{w}_0}$ a.s., as needed. ☐

Theorem 7.13. *Let D be a convex $C^{1,\alpha}$ domain $(0 < \alpha < 1)$ which is symmetric with respect to the horizontal axis. If φ is a second Neumann eigenfunction for D which is antisymmetric with respect to the horizontal axis, then φ is monotone on the family of hyperbolic lines in D which intersect the horizontal axis.*

In particular, φ must attain its maximum and minimum over $\overline{D}$ on the boundary of D.

Proof. By assumption, φ must be identically zero on the horizontal axis, and therefore the nodal line for φ is Γ_0 (the part of the horizontal axis contained in $\overline{D}$). It follows that D^+ and D^- are the nodal domains of φ.

Since φ has constant sign on each nodal domain, without loss of generality we will assume that φ is positive on D^+. Since φ is antisymmetric with respect to the horizontal axis, it suffices to prove the monotonicity of φ in D^+ along the indicated family of curves.

Consider any hyperbolic line in D, which intersects the horizontal axis, and denote by w_1 the point of intersection. If $w_0, \tilde{w}_0 \in \overline{D^+}$ are arbitrarily chosen points lying on this hyperbolic line, such that $w_0 \in [w_1, \tilde{w}_0]$, we will show that $\varphi(w_0) \leq \varphi(\tilde{w}_0)$.

Since the restriction of a second Neumann eigenfunction for D to one of its nodal domains has constant sign, it follows that it is the first mixed Dirichlet-Neumann eigenfunction for the corresponding nodal domain. Therefore, the restriction of φ to $\overline{D^+}$ is the first mixed Dirichlet-Neumann eigenfunction for D^+, with Neumann conditions on Γ^+ and Dirichlet conditions on Γ_0.

It can be shown that the transition density $p_{D^+}(t, x, y)$ of reflected Brownian motion in D^+, killed on hitting the horizontal axis, has an eigenfunction expansion in terms of the mixed Dirichlet-Neumann eigenfunctions for D^+, with Dirichlet boundary conditions on Γ_0 and Neumann conditions on Γ^+. More precisely, it can be shown that:

$$p_{D^+}(t, x, y) = \sum_{i \geq 1} e^{-\mu_i t} \varphi_i(x) \varphi_i(y),$$

where $0 < \mu_1 < \mu_2 \leq \dots$ are the mixed Dirichlet-Neumann eigenvalues for D^+ repeated according to the multiplicity, and $\{\varphi_i\}_{i \geq 1}$ is an orthonormal sequence of eigenfunctions corresponding to the eigenvalues $\{\mu_i\}_{i \geq 1}$. The convergence is uniform and absolute on $\overline{D^+}$.

Note that since μ_1 is simple, the corresponding eigenspace is one-dimensional, and therefore $\varphi_1 = c\varphi$, for some nonzero constant c. Also note that since $\mu_1 < \mu_i$, for all $i \geq 2$, we can write:

$$p_{D^+}(t, x, y) = e^{-\mu_1 t} \varphi_1(x) \varphi_1(y) + R(t, x, y) \qquad (7.13)$$
$$= c^2 e^{-\mu_1 t} \varphi(x) \varphi(y) + R(t, x, y),$$

where $\lim\limits_{t \to \infty} e^{\mu_1 t} R(t, x, y) = 0$, uniformly in $x, y \in \overline{D^+}$.

Consider the function $u : (0, \infty) \times D \to \mathbb{R}$ given by $u(t, x) = \mathbb{E}[1; \tau^x > t]$, where τ^x is the lifetime of reflected Brownian motion in D^+ starting at x, killed on hitting Γ_0. Integrating the eigenfunction expansion (7.13), we obtain:

$$u(t, x) = \int_{D^+} p_{D^+}(t, x, y) dy \qquad (7.14)$$

$$= c^2 e^{-\mu_1 t} \varphi(x) \int_{D^+} \varphi(y) dy + \int_{D^+} R(t, x, y) dy$$

$$= a e^{-\mu_1 t} \varphi(x) + R_1(t, x),$$

where, by assumption, $a = c^2 \int_{D^+} \varphi(y) dy > 0$ and $R_1(t, x)$ approaches zero faster than $e^{-\mu_1 t}$ as $t \to \infty$.

Since $u(t, x) = \mathbb{P}(\tau^x > t)$, and using the monotonicity property in Lemma 7.12, it follows that for any $t \geq 0$ we have:

$$u(t, w_0) = \mathbb{P}(\tau^{w_0} > t) \leq \mathbb{P}(\tau^{\tilde{w}_0} > t) = u(t, \tilde{w}_0).$$

Using the analytic representation (7.14) of $u(t, \cdot)$ we obtain therefore:

$$a\varphi(w_0) + e^{\mu_1 t} R_1(t, w_0) \leq a\varphi(\tilde{w}_0) + e^{\mu_1 t} R_1(t, \tilde{w}_0),$$

for all $t \geq 0$. Letting $t \to \infty$, it follows $a\varphi(w_0) \leq a\varphi(\tilde{w}_0)$, and since $a > 0$, we obtain $\varphi(w_0) \leq \varphi(\tilde{w}_0)$, as needed. $\square$

Chapter 8
Nodal Lines

This chapter is based on [AB02]. Suppose that D is a planar domain (open and connected set) with a piecewise smooth boundary, i.e., the boundary consists of finitely many C^2 parts, and the uniform exterior sphere condition and the uniform interior cone condition are satisfied. Under these assumptions on D, one can prove strong existence and pathwise uniqueness for reflected Brownian motion in $\overline{D}$ (see [LS84]).

Recall that the first eigenvalue for the Laplacian in D with Neumann boundary conditions is equal to 0 and the corresponding eigenfunction is constant. The second eigenvalue need not be simple but its multiplicity can be only 1 or 2 [Nad86, Nad87, NTY01]. The nodal set is the set of points in D where the second eigenfunction vanishes. If D is not simply connected, the nodal set need not be a connected curve. The nodal set of any second Neumann eigenfunction divides the domain D into exactly two nodal domains (connected and open sets).

Let x, y be points and K be a straight line, and denote by A_1 and A_2 the two connected components of $\mathbb{R}^2 \setminus K$. We say that x and y lie on the opposite sides of K, if either $x \in \overline{A_1}$ and $y \in \overline{A_2}$, or $x \in \overline{A_2}$ and $y \in \overline{A_1}$.

Theorem 8.1. *Suppose that $A \subset \overline{D}$ is closed, $x \in D \setminus A$, D_1 is a connected component of $D \setminus A$ which contains x, and $y \in D \setminus D_1$. Assume that there exists a coupling (X_t, Y_t) of reflected Brownian motions in $\overline{D}$ and a line-valued stochastic process K_t with the following properties. For every $t \geq 0$, X_t and Y_t lie on the opposite sides of K_t. Moreover, $(X_0, Y_0) = (x, y)$ and $K_t \cap D \subset A$ for all $t \geq 0$ a.s. Then for any second Neumann eigenfunction in D, none if its nodal domains can have a closure which is a subset of $\overline{D} \setminus A$ containing x.*

Proof. Suppose that F is one of the nodal domains for a second eigenfunction and $x \in \overline{F} \subset \overline{D} \setminus A$. We will show that this assumption leads to a contradiction. Let C_1 be a non-empty open disc centered at a boundary point of F, but not on the boundary of D, so small that $F_* = F \cup C_1$ satisfies $x \in \overline{F}_* \subset \overline{D} \setminus A$, and there exists a non-empty open disc $C_2 \subset D \setminus F_*$. Assume that y, X_t, Y_t and K_t satisfy the assumptions of the theorem. Since the stationary measure for X_t is the uniform distribution in D,

K. Burdzy, *Brownian Motion and its Applications to Mathematical Analysis*,
Lecture Notes in Mathematics 2106, DOI 10.1007/978-3-319-04394-4_8,
© Springer International Publishing Switzerland 2014

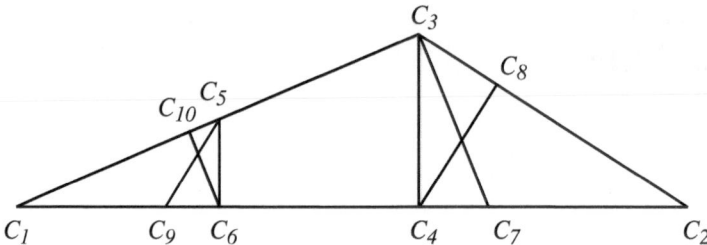

Fig. 8.1 *Obtuse triangle* and its subsets indicating the position of the second Neumann eigenfunction nodal line

X_t will hit C_2 with probability one, and so it will leave F_* with probability 1. Recall that there are exactly two nodal domains—F is one of them; let F_1 denote the other one. Let $\tau_X(F_*) = \inf\{t \geq 0 : X_t \notin F_*\}$ and $\tau_Y(F_1) = \inf\{t \geq 0 : Y_t \notin F_1\}$. We will argue that $\tau_X(F_*) < \tau_Y(F_1)$ a.s. Suppose otherwise. By the continuity of $t \rightarrow Y_t$, $Y_{\tau_Y(F_1)}$ belongs to $\partial F_1 \setminus \partial D$, and so $Y_{\tau_Y(F_1)} \in \overline{F} \subset \overline{F}_*$. We have assumed that $\tau_X(F_*) \geq \tau_Y(F_1)$, so $X_{\tau_Y(F_1)} \in \overline{F}_*$. Since the points $X_{\tau_Y(F_1)}$ and $Y_{\tau_Y(F_1)}$ belong to the connected set $\overline{F}_*$, the line $K_{\tau_Y(F_1)}$ must also intersect $\overline{F}_*$. We have $K_t \cap D \subset A$ for all t so $A \cap \overline{F}_* \neq \emptyset$, which is a contradiction. We conclude that $\tau_X(F_*) < \tau_Y(F_1)$ a.s.

Let $\mu > 0$ denote the second Neumann eigenvalue for D (recall that the first eigenvalue is equal to 0). Then μ is the first eigenvalue for the mixed problem in the nodal domain F, with the Neumann boundary conditions on ∂D and the Dirichlet boundary conditions on the nodal line. The fact that F_* is strictly larger than F easily implies that $\mu > \mu^*$, where μ^* is the analogous mixed eigenvalue for F_*. Using the well known identification of Brownian motion density with the heat equation solution, we obtain from [BB99, Proposition 2.1] that

$$\lim_{t \to \infty} \mathbb{P}(\tau_Y(F_1) > t)e^{\mu t} \in (0, \infty).$$

Since $\tau_X(F_*) < \tau_Y(F_1)$ a.s.,

$$\lim_{t \to \infty} \mathbb{P}(\tau_X(F_*) > t)e^{\mu t} < \infty,$$

and so $\mu^* \geq \mu$, but this contradicts the fact that $\mu > \mu^*$. Our initial assumption that $x \in \overline{F} \subset \overline{D} \setminus A$ for some nodal domain F must be false. $\square$

Homework 8.2. Use mirror couplings and Theorem 8.1 to show that in the obtuse triangle depicted in Fig. 8.1, the nodal line has to intersect the subset of D which lies to the right of the circular arc passing through C_5 and centered at C_1. See Example 8.5 for the explanation of notation in Fig. 8.1.

We will now introduce a new coupling which gives stronger results than the mirror coupling for some families of domains. First, we will define a family of domains.

Let β denote the mapping which assigns to each point in $\mathbb{R}^2$ its symmetric image about the vertical axis. Let (e_1, e_2) be the usual orthonormal basis for $\mathbb{R}^2$ and let $\mathbf{U}_+$ be the right half plane, i.e., $\mathbf{U}_+ = \{x \in \mathbb{R}^2 : x \cdot e_1 \geq 0\}$. We will denote the left half plane $\mathbf{U}_- = \beta \mathbf{U}_+$. A convex set $C \subset \mathbb{R}^2$ will be called a cone if $\alpha x \in C$ for any $x \in C$ and $\alpha \geq 0$.

Recall that D is a planar piecewise smooth domain. We will define a family of domains via conditions (A1)–(A4) below. Conditions (A3)–(A4) are the most important of the four conditions because they impose considerable restrictions on the shape of D. Assume that D intersects the vertical axis. Let

$$D_1 = D \cap \mathbf{U}_+, \quad D_2 = D \cap \mathbf{U}_-,$$

$$\tilde{D} = \beta D, \quad \tilde{D}_1 = \beta D_1, \quad \tilde{D}_2 = \beta D_2.$$

For $i = 1, 2$, let ∂D_i denote the set of all boundary points of D_i, where the boundary is smooth, except those that lie on the vertical axis. Let $\partial \tilde{D}_i$ be defined similarly. We will write $\mathbf{n}(x)$ for the inward normal to ∂D_i at $x \in \partial D_i$, for $i = 1, 2$. With an abuse of notation, the same symbol $\mathbf{n}(x)$ will be used to denote the inward normal to $\partial \tilde{D}_i$ at $x \in \partial \tilde{D}_i$.

(A1) $D_1 \subset \tilde{D}_2$.
(A2) For all $x \in \partial D \cap \partial \tilde{D}$ except for a finite set of such x, the sets ∂D and $\partial \tilde{D}$ agree in some neighborhood of x.
(A3) There exists a closed cone $C \subset \mathbf{U}_-$ such that for any $x \in \partial D_1 \setminus \partial \tilde{D}_2$, one has $\mathbf{n}(x) \in C$.

In the case when C is a proper subset of $\mathbf{U}_-$, let v_1 and v_2 denote the two unit vectors that generate the cone C, i.e., all vectors in C are linear combinations of v_1 and v_2 with non-negative coefficients. For $i = 1, 2$, let $v_i^\perp$ denote the unit vector orthogonal to v_i, such that $v_i^\perp \cdot v_{3-i} > 0$. In the case when $C = \mathbf{U}_-$, let $v_1 = -e_2$, $v_2 = e_2$, and $v_1^\perp = v_2^\perp = -e_1$. Note that in both cases,

$$v_i + \varepsilon v_i^\perp \in C, \tag{8.1}$$

provided that $\varepsilon > 0$ is small enough.

For $x \in \mathbb{R}^2$, let the ray through x be denoted as $R(x) = \{\alpha x : \alpha > 0\}$.
(A4) (i) For $y \in \partial \tilde{D}_2 \cup (\partial \tilde{D}_1 \cap \partial D_2)$ and $i = 1, 2$, if $y + R(v_i)$ intersects $\overline{D}$ then $\mathbf{n}(x) \cdot v_i^\perp \leq 0$ for all $x \in \partial \tilde{D}_2$ in a neighborhood of y. (ii) For $y \in \partial D_2 \cup (\partial D_1 \cap \partial \tilde{D}_2)$ and $i = 1, 2$, if $y - R(v_i)$ intersects $\overline{\tilde{D}}$ then $\mathbf{n}(x) \cdot v_i^\perp \geq 0$ for all $x \in \partial D_2$ in a neighborhood of y.

Suppose that $x \in D \cap \tilde{D}$ and W_t is a planar Brownian motion. Then there exist a reflected Brownian motion X_t in D, and a reflected Brownian motion Y_t in $\tilde{D}$, driven by W_t, and starting from x:

$$X_t = x + W_t + \int_0^t \mathbf{n}(X_s)dL_s, \quad Y_t = x + W_t + \int_0^t \mathbf{n}(Y_s)dM_s. \quad (8.2)$$

Here L_t and M_t are local times on the boundary for X_t and Y_t. Note that X and Y are driven by the same Brownian motion but the pair of process (X, Y) does not form a "synchronous coupling" in the sense of Chap. 5 because X and Y have different state spaces.

Theorem 8.3. *Assume that D satisfies conditions (A1)–(A4) and X_t and Y_t satisfy (8.2). Then with probability one, for all $t \geq 0$, $X_t - Y_t \in C$.*

Proof. We have assumed that ∂D is smooth, except for a finite number of points—it is well known that the set of these points is polar for the reflected Brownian motion in D. For all $x \in \partial D$ outside this set, $\mathbf{n}(x)$ is well defined and $\mathbf{n}(y)$ depends continuously on $y \in \partial D$ in a neighborhood of x. Assume that the assertion of the theorem does not hold, and let $\sigma = \inf\{t : X_t - Y_t \notin C\}$. Clearly, $X_\sigma - Y_\sigma \in \partial C$.

If $X_\sigma \in D$ and $Y_\sigma \in \tilde{D}$, then there is $\varepsilon > 0$ such that $X_t \in D$ and $Y_t \in \tilde{D}$ for all $t \in [\sigma, \sigma + \varepsilon]$. By (8.2), $X_t - Y_t = X_\sigma - Y_\sigma \in \partial C$ for all such t, contradicting the definition of σ. Hence either $X_\sigma \in \partial D$ or $Y_\sigma \in \partial \tilde{D}$.

If $X_\sigma = Y_\sigma$, then using condition (A1), either $X_\sigma \in \partial D_1$ or $Y_\sigma \in \partial \tilde{D}_1$. Since the cases are similar, we will discuss only the former. Consider the following two cases.

(a) If $X_\sigma \in \partial D \cap \partial \tilde{D}$ then, by (A2), ∂D and $\partial \tilde{D}$ agree in a small neighborhood of X_σ. Using (8.2) and strong uniqueness, we see that $X_t = Y_t$ for $t \in [\sigma, \sigma + \delta]$, if $\delta > 0$ is sufficiently small. This contradicts the definition of σ.
(b) The other case is when $X_\sigma \in \partial D_1 \setminus \partial \tilde{D}_2$ and $Y_\sigma \in \tilde{D}$. Then for small $\varepsilon > 0$ and all $t \in [\sigma, \sigma + \varepsilon]$, $X_t - Y_t = \int_\sigma^t \mathbf{n}(X_s)dL_s$. Condition (A3) implies that $\int_\sigma^t \mathbf{n}(X_s)dL_s \in C$ for all such t, provided that ε is small enough, contradicting the definition of σ.

We conclude that $X_\sigma \neq Y_\sigma$ and so $X_\sigma - Y_\sigma \in \partial C \setminus \{0\}$. In other words,

$$X_\sigma - Y_\sigma = \alpha v_i, \quad (8.3)$$

where $\alpha > 0$ and either $i = 1$ or $i = 2$.

The rest of the argument is split into three cases with subcases.

Case 1: $Y_\sigma \in \tilde{D}$.

(a) $X_\sigma \in \partial D_1 \setminus \partial \tilde{D}_2$. In this case, $X_t - Y_t = \alpha v_i + \int_\sigma^t \mathbf{n}(X_s)dL_s$ for $t \in [\sigma, \sigma + \varepsilon]$ provided that $\varepsilon > 0$ is small enough. By condition (A3), $X_t - Y_t \in C$ for all such t, taking, if necessary, $\varepsilon > 0$ smaller. This contradicts the definition of σ.

(b) $X_\sigma \in \partial D_2 \cup (\partial D_1 \cap \partial \tilde{D}_2)$. This is similar to Case 2 (a) considered below, except that we should use condition (A4) (ii) in place of (A4) (i).

Case 2: $Y_\sigma \in \partial \tilde{D}_2 \cup (\partial \tilde{D}_1 \cap \partial D_2)$.

A single argument will apply to subcases (a) and (b):

(a) $X_\sigma \in D$.

(b) $X_\sigma \in \partial D_1 \setminus \partial \tilde{D}_2$.

In either case we have

$$X_t - Y_t = \alpha v_i + \int_\sigma^t \mathbf{n}(X_s)dL_s - \int_\sigma^t \mathbf{n}(Y_s)dM_s. \tag{8.4}$$

Since $X_\sigma \in \overline{D}$, (8.3) implies that $Y_\sigma + R(v_i)$ intersects $\overline{D}$. Hence, by condition (A4) (i), $\int_\sigma^t \mathbf{n}(Y_s)dM_s \cdot v_i^\perp \leq 0$ for $t \in [\sigma, \sigma + \varepsilon]$, provided $\varepsilon > 0$ is small. As before, condition (A3) implies that $\int_\sigma^t \mathbf{n}(X_s)dL_s \in C$. As a result, using (8.1), $X_t - Y_t \in C$ for $t \in [\sigma, \sigma + \varepsilon]$ if $\varepsilon > 0$ is small enough. This contradicts the definition of σ.

(c) $X_\sigma \in \partial D_2 \cup (\partial D_1 \cap \partial \tilde{D}_2)$. In this case (8.4) again holds, and using condition (A4) (i), $\int_\sigma^t \mathbf{n}(Y_s)dM_s \cdot v_i^\perp \leq 0$. Moreover, by condition (A4) (ii), $\int_\sigma^t \mathbf{n}(X_s)dL_s \cdot v_i^\perp \geq 0$. An application of (8.1) again shows that $X_t - Y_t \in C$ for $t \in [\sigma, \sigma + \varepsilon]$ and $\varepsilon > 0$ small. This is impossible, in view of the definition of σ.

Case 3: $Y_\sigma \in \partial \tilde{D}_1 \setminus \partial D_2$.

(a) $X_\sigma \in D$. Note that condition (A3) implies that for $y \in \partial \tilde{D}_1 \setminus \partial D_2$, $\mathbf{n}(y) \in -C$. Hence this case is similar to Case 1 (a), with the difference that now we have $X_t - Y_t = \alpha v_i - \int_\sigma^t \mathbf{n}(Y_s)dM_s$ for $t \in [\sigma, \sigma + \varepsilon]$.

(b) $X \in \partial D_1 \setminus \partial \tilde{D}_2$. Since $\partial \tilde{D}_1 + C$ does not intersect ∂D_1, this case is ruled out by (8.3).

(c) $X_\sigma \in \partial D_2 \cup (\partial D_1 \cap \partial \tilde{D}_2)$. This case is similar to Case 2 (b) considered above.

We arrived at a contradiction in all cases. As a result, $X_t - Y_t \in C$ must hold for all $t \geq 0$, with probability one.   □

Recall that e_1 is the first base vector of $\mathbb{R}^2$. Under the assumptions of Theorem 8.3 we have $e_1 \cdot (X_t - Y_t) \leq 0$ for all t. Let $\tilde{Y} = \beta Y$ and note that X_t and $\tilde{Y}_t$ are reflected Brownian motions in D. Then $e_1 \cdot (X_t + \tilde{Y}_t) \leq 0$ for all t. This implies that if K_t denotes the vertical line passing through the midpoint of X_t and $\tilde{Y}_t$ then $K_t \subset U_-$ for all $t \geq 0$. This and Theorem 8.1, applied with $A = U_- \cap \overline{D}$, yield the following.

Corollary 8.4. *Assume that D satisfies conditions (A1)–(A4). Then for any second Neumann eigenfunction in D, its nodal line must intersect $\overline{D}_2$.*

We will present two examples illustrating Corollary 8.4.

Example 8.5. Suppose that D is an obtuse triangle with vertices C_1, C_2 and C_3. Let C_3 be the vertex with an angle greater than $\pi/2$, and suppose that the angle at C_2 is not smaller than that at C_1 (see Fig. 8.1). Let $\overline{C_j C_k}$ denote the line segment with endpoints C_j and C_k. The points C_k, $k = 4, \ldots, 10$, are chosen so that $C_5, C_{10} \in \overline{C_1C_3}$, $C_8 \in \overline{C_2C_3}$, $C_9, C_6, C_4, C_7 \in \overline{C_1C_2}$, and the following pairs of line segments are perpendicular: $\overline{C_1C_2}$ and $\overline{C_3C_4}$, $\overline{C_1C_2}$ and $\overline{C_5C_6}$, $\overline{C_2C_3}$ and $\overline{C_5C_9}$, $\overline{C_2C_3}$ and $\overline{C_4C_8}$, $\overline{C_1C_3}$ and $\overline{C_7C_3}$, $\overline{C_1C_3}$ and $\overline{C_6C_{10}}$. The point C_6 lies half way between C_1 and C_4.

Let A be the closed subset of $\overline{D}$ (trapezoid) with vertices C_3, C_4, C_6 and C_5. Let A_1 be the closure of the union of the quadrilateral with vertices $C_3C_7C_9C_5$ and pentagon $C_3C_8C_4C_6C_{10}$.

The second Neumann eigenvalue is simple in obtuse triangles by a theorem in [AB04].

We will show that

(i) The nodal line for the second Neumann eigenfunction must intersect A.
(ii) The nodal line lies within A_1.

First, place the coordinate system so that the vertical axis passes through $\overline{C_3C_4}$. Then let v_1 be the vector perpendicular to $\overline{C_2C_3}$, pointing to the left, and let v_2 be its image under the map $(x_1, x_2) \mapsto (x_1, -x_2)$.

Homework 8.6. Verify assumptions (A1)–(A4).

We apply Corollary 8.4 to conclude that the nodal line must intersect the closed triangle $C_1C_4C_3$.

Next place the coordinate system in such a way that $\overline{C_5C_6}$ lies on the vertical axis. Then flip the triangle around the vertical axis, i.e., apply the transformation we call β. We will apply Theorem 8.3 and Corollary 8.4 to this new triangle. In this case we take $C = \mathbf{U}_-$.

Homework 8.7. Verify assumptions (A1)–(A4).

We conclude that the nodal line must cross the closed quadrilateral $C_6C_2C_3C_5$. This completes the proof of (i).

Claim (ii) is a consequence of (i) and the bounds on the direction of the gradient of the second eigenfunction proved in [BB99, Theorem 3.1].

Remark 8.8. We will argue heuristically that it is impossible to obtain much sharper estimates for the nodal line location. It is not hard to show that if the triangle is isosceles then the nodal line is the line of symmetry. This shows that $\overline{C_3C_4}$ is a sharp "bound". To see how sharp is the other "bound", i.e., $\overline{C_5C_6}$, consider an obtuse triangle with two angles very close to $\pi/2$—one slightly less than the right angle and another one slightly larger than that (see Fig. 8.2). Such a triangle has a shape very close to a thin circular sector. Let us assume that the diameter of the triangle is equal to 1. The nodal line in a circular sector is an arc with center at its vertex A. The nodal line distance from A is equal to $a_0/a_1 \approx 0.63$, where a_0 is the first positive zero of the Bessel function of order 0 and a_1 is the first positive zero of its derivative. This follows from known results on eigenfunctions in discs and estimates for Bessel function zeroes, see [Ban80, p. 92]. Our methods yield 0.5 as the lower bound for

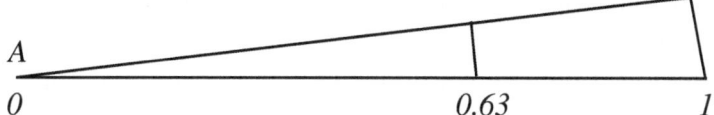

Fig. 8.2 An *obtuse triangle* with two angles very close to $\pi/2$—one slightly less than the right angle and another one slightly larger than that

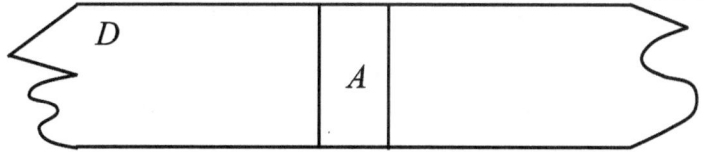

Fig. 8.3 A "rectangle-like" domain and a set that must be intersected by the second Neumann eigenfunction nodal line

the distance of the nodal line from A; we see that this estimate cannot be improved beyond 0.63.

Conjecture 8.9. The nodal line for the second Neumann eigenfunction is contained in the triangle $C_1 C_3 C_4$.

Little is known about eigenfunctions in triangles different from equilateral; see [Pin80, Pin85] for that special case. For some exciting recent progress, see [Pol] (do not miss links to earlier threads of this project).

Example 8.10. Consider a domain D in the plane whose boundary contains line segments $\{(x_1, x_2) : -a \leq x_1 \leq a, x_2 = 0\}$ and $\{(x_1, x_2) : -a \leq x_1 \leq a, x_2 = 1\}$ for some $a > 0$. Suppose that the remaining parts of ∂D are contained in $\{(x_1, x_2) : -a - b \leq x_1 \leq -a, 0 \leq x_2 \leq 1\}$ and $\{(x_1, x_2) : a \leq x_1 \leq a + b, 0 \leq x_2 \leq 1\}$, and moreover, they are graphs of functions in the coordinate system with the basis $(e_2, -e_1)$, i.e., the usual coordinate system rotated by the angle $\pi/2$. Let

$$A = \{(x_1, x_2) \in \overline{D} : -b/2 \leq x_1 \leq b/2\},$$

$$A_1 = \{(x_1, x_2) \in \overline{D} : -b/2 - 1 \leq x_1 \leq b/2 + 1\}.$$

Figure 8.3 shows an example of D and the corresponding rectangle A.
  We will prove that

(i) The nodal line for any second Neumann eigenfunction must intersect A.
     Recall that we call a set D a "lip domain" if its boundary consists of graphs of two Lipschitz functions with the Lipschitz constant 1. The second eigenvalue is simple in every lip domain except square by Theorem 6.8.
     Our second claim is

(ii) If D is a lip domain then the nodal line for the Neumann eigenfunction lies within A_1.

Place the coordinate system so that the right vertical edge of the boundary of A lies on the vertical axis. Let the cone C be equal to U_-.

Homework 8.11. Verify assumptions (A1)–(A4).

By Theorem 8.3 and Corollary 8.4, it follows that the nodal line has to intersect the set to the left of the right edge of A. A similar argument applies to the left edge of A and this completes the proof of (i).

The second claim follows from the first one and the bounds on the direction of the gradient of the second eigenfunction in lip domains given in [BB99, Example 3.1].

Chapter 9
Neumann Heat Kernel Monotonicity

This chapter is based on a paper by Pascu and Gageonea [PG11] and is devoted to the proof of the "Laugesen-Morpurgo conjecture," that is, Theorem 9.1 below (see [PG11] for the history of the problem).

We denote by $\mathbb{U} = \{z \in \mathbb{R}^n : \|z\| < 1\}$ the open unit ball in $\mathbb{R}^n$ ($n \geq 1$). Let $p_{\mathbb{U}}(t, x, y)$ denote the heat kernel for the Laplacian with Neumann boundary conditions on $\mathbb{U}$ (or equivalently, the transition density of reflected Brownian motion in $\mathbb{U}$).

Theorem 9.1. *Let $p_{\mathbb{U}}(t, x, y)$ denote the heat kernel for the Laplacian with Neumann boundary conditions on the unit ball $\mathbb{U} = \{z \in \mathbb{R}^n : \|z\| < 1\}$ in $\mathbb{R}^n$ ($n \geq 1$).*
For any $t > 0$ we have

$$p_{\mathbb{U}}(t, x, x) < p_{\mathbb{U}}(t, y, y), \tag{9.1}$$

for all $x, y \in \overline{\mathbb{U}}$ with $\|x\| < \|y\|$.

The proof will be preceded by some preliminary results. Given a hyperplane $\mathcal{H} \subset \mathbb{R}^n$, we say that the points $x, y \in \mathbb{R}^n$ are separated by $\mathcal{H}$ if x and y lie in different components of $\mathbb{R}^n - \mathcal{H}$, and we say that they are not separated by $\mathcal{H}$ otherwise.

Recall the mirror coupling. Given any points $x, y \in \overline{\mathbb{U}}$, we define the mirror coupling of reflected Brownian motions in the unit ball $\mathbb{U} \subset \mathbb{R}^n$ as a pair $(X_t, Y_t)_{t \geq 0}$ of stochastic processes given by

$$\begin{cases} X_t = x + W_t + \int_0^t \mathbf{n}(X_s) \, dL_s^X \\ Y_t = y + Z_t + \int_0^t \mathbf{n}(Y_s) \, dL_s^Y \end{cases}, \tag{9.2}$$

where $\mathbf{n}(x)$ is the unit inward vector at $x \in \partial \mathbb{U}$, W_t is a n-dimensional Brownian motion starting at $W_0 = 0$, Z_t is the mirror image of the Brownian motion W_t with respect to the hyperplane $\mathcal{M}_t$ of symmetry between X_t and Y_t, that is

K. Burdzy, *Brownian Motion and its Applications to Mathematical Analysis*,
Lecture Notes in Mathematics 2106, DOI 10.1007/978-3-319-04394-4_9,
© Springer International Publishing Switzerland 2014

$$Z_t = W_t - 2 \int_0^t \frac{X_s - Y_s}{\|X_s - Y_s\|^2} (X_s - Y_s) \cdot dW_s, \tag{9.3}$$

and L_t^X and L_t^Y denote the boundary local times of the reflected Brownian motions X_t and respectively Y_t. The processes X_t and Y_t evolve according to (9.2) above for $t \leq \tau$, where τ is the coupling time

$$\tau = \inf \{t > 0 : X_t = Y_t\} \in \mathbb{R} \cup \{\infty\},$$

and they evolve together after the coupling time (i.e. $X_t = Y_t$ for $t \geq \tau$).

The key to proving the Laugesen-Morpurgo conjecture is the double inequality (9.13) in Theorem 7.13, which in turn relies on proving the following inequality:

$$p_{\mathbb{U}}(t, y, z) \leq p_{\mathbb{U}}(t, x, z), \qquad t > 0, \tag{9.4}$$

for all $x, y, z \in \mathbb{U}$ satisfying $\|x - z\| \leq \|y - z\|$ and $\|y\| \leq \|x\|$.

Consider a mirror coupling X_t, Y_t of reflected Brownian motions in $\mathbb{U}$ given by (9.2)–(9.3), with starting points $X_0 = x$, $Y_0 = y \in \overline{\mathbb{U}}$.

For $t < \tau = \inf\{t > 0 : X_t = Y_t\}$, the mirror $\mathcal{M}_t$ of the coupling (the hyperplane of symmetry between X_t and Y_t) is given by

$$\mathcal{M}_t = \left\{ z \in \mathbb{R}^n : \left(z - \frac{X_t + Y_t}{2} \right) \cdot (X_t - Y_t) = 0 \right\}. \tag{9.5}$$

The idea for proving the inequality (9.4) is that the mirror $\mathcal{M}_t$ moves towards the origin, in the sense of Lemma 9.2 below. This property is a rigorous version of [BK00, Example 4.5], used to prove the efficiency of the mirror coupling in the case of the unit disk.

Lemma 9.2. *Let X_t, Y_t be a mirror coupling of reflected Brownian motions in $\mathbb{U}$ with starting points $X_0 = x$, $Y_0 = y \in \overline{\mathbb{U}}$, and let $\tau = \inf\{t > 0 : X_t = Y_t\}$ be the coupling time and $\tau_1 = \inf\{t > 0 : 0 \in \mathcal{M}_t\}$.*

For all times $t < \tau \wedge \tau_1$, the mirror $\mathcal{M}_t$ moves towards the origin, in such a way that if a point $R \in \mathbb{U}$ and the origin are separated by $\mathcal{M}_{t_1}$ for some $t_1 \in [0, \tau \wedge \tau_1)$, then the point R and the origin are separated by $\mathcal{M}_{t_2}$ for all $t_2 \in [t_1, \tau \wedge \tau_1)$ (see Figs. 9.1 and 9.2).

Proof. If $\|x\| = \|y\|$, then $\tau_1 = 0$ and there is nothing to prove in this case (the mirror $\mathcal{M}_0$ passes through the origin). Without loss of generality we may therefore assume that $\|x\| > \|y\|$.

Setting

$$\begin{cases} U_t = X_t - Y_t, \\ V_t = X_t + Y_t, \end{cases} \qquad t \geq 0, \tag{9.6}$$

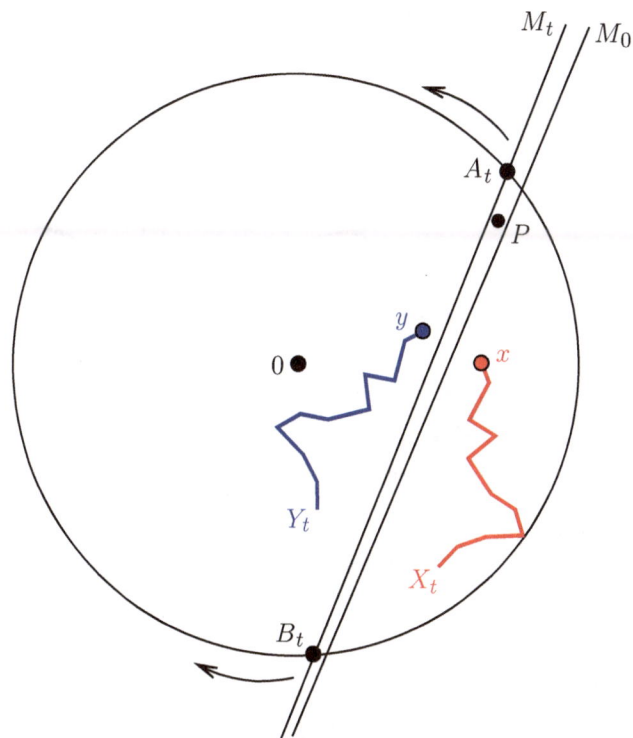

Fig. 9.1 Mirror coupling of reflected Brownian motions in the unit disk (figure from [PG11])

from the definition (9.2)–(9.3) of the mirror coupling we obtain:

$$\begin{cases} U_t^i = x^i - y^i + W_t^i - Z_t^i - \int_0^t X_s^i dL_s^X + \int_0^t Y_s^i dL_s^Y \\ V_t^i = x^i + y^i + W_t^i + Z_t^i - \int_0^t X_s^i dL_s^X - \int_0^t Y_s^i dL_s^Y \end{cases}, \quad i = 1, \ldots, n,$$

for all $t \leq \tau$, where the superscript i indicates the ith cartesian coordinate of the given point.

Using the definition (9.3) of Z_t, we have

$$\begin{cases} U_t^i = x^i - y^i + 2\int_0^t \frac{U_s^i}{\|U_s\|^2} U_s \cdot dW_s - \int_0^t X_s^i dL_s^X + \int_0^t Y_s^i dL_s^Y \\ V_t^i = x^i + y^i + 2W_t^i - 2\int_0^t \frac{U_s^i}{\|U_s\|^2} U_s \cdot dW_s - \int_0^t X_s^i dL_s^X - \int_0^t Y_s^i dL_s^Y \end{cases}, \quad (9.7)$$

for all $i = 1, \ldots, n$ and $t < \tau$, and therefore we obtain the following formulae for the quadratic variation of the processes U and V:

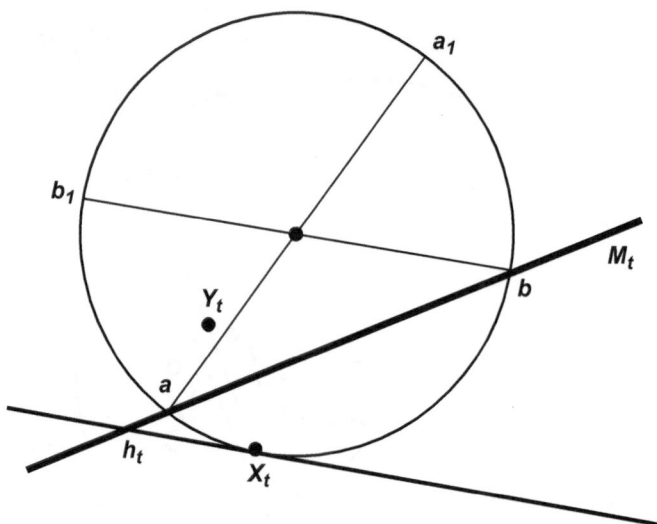

Fig. 9.2 Mirror coupling of reflected Brownian motions in the unit disk (figure from [BK00])

$$
\begin{cases}
\langle U^i, U^j \rangle_t = 4 \int_0^t \frac{U_s^i U_s^j}{\|U_s\|^2} \, ds \\
\langle V^i, V^j \rangle_t = 4 \int_0^t \delta_{ij} - \frac{U_s^i U_s^j}{\|U_s\|^2} \, ds \;, \qquad i, j, = 1, \ldots, n. \qquad (9.8) \\
\langle U^i, V^j \rangle_t = 0
\end{cases}
$$

Note that since $\|X_0\| = \|x\| > \|y\| = \|Y_0\|$, it follows that for all $t < \tau \wedge \tau_1$ we have

$$
U_t \cdot V_t = (X_t - Y_t) \cdot (X_t + Y_t) = \|X_t\|^2 - \|Y_t\|^2 > 0, \qquad (9.9)
$$

and therefore for $t < \tau \wedge \tau_1$ we may define the process A_t by

$$
A_t = \frac{2}{U_t \cdot V_t} U_t. \qquad (9.10)
$$

We will first show that for $t < \tau \wedge \tau_1$ the components of the process A_t are processes of bounded variation, satisfying

$$
dA_t^i = \frac{2}{U_t \cdot V_t} \left(A_t^i - \frac{U_t^i + V_t^i}{2} \right) dL_t^X, \qquad i = 1, \ldots, n. \qquad (9.11)
$$

Applying the Itô formula to the C^2 function $f(u, v) = u^i / (u \cdot v)$ and to the processes U_t and V_t, we have:

$$\frac{1}{2}dA_t^i = d\left(\frac{U_t^i}{U_t \cdot V_t}\right)$$

$$= \frac{1}{(U_t \cdot V_t)^2} \sum_{j=1}^{n} \left(\left(\delta_{ij}\, U_t \cdot V_t - U_t^i V_t^j\right) dU_t^j - U_t^i U_t^j dV_t^j\right)$$

$$+ \frac{1}{2(U_t \cdot V_t)^3} \sum_{j,k=1}^{n} \left(2U_t^i V_t^j V_t^k - \delta_{ij} V_t^k U_t \cdot V_t - \delta_{ik} V_t^j U_t \cdot V_t\right) d\langle U^j, U^k\rangle_t$$

$$+ \frac{1}{2(U_t \cdot V_t)^3} \sum_{j,k=1}^{n} \left(2U_t^i U_t^j U_t^k\right) d\langle V^j, V^k\rangle_t$$

$$+ \frac{1}{(U_t \cdot V_t)^3} \sum_{j,k=1}^{n} \left(2U_t^i U_t^k V_t^j - \delta_{ij} U_t^k U_t \cdot V_t - \delta_{jk} U_t^i U_t \cdot V_t\right) d\langle U^j, V^k\rangle_t.$$

Using the relations in (9.7) it can be seen that the martingale part in the above expression reduces to zero, and combining with (9.8) we obtain

$$\frac{1}{2}dA_t^i = \frac{1}{(U_t \cdot V_t)^2} \sum_{j=1}^{n} \left(\left(\delta_{ij}\, U_t \cdot V_t - U_t^i V_t^j\right)\left(-X_t^j dL_t^X + Y_t^j dL_t^Y\right)\right)$$

$$- \frac{1}{(U_t \cdot V_t)^2} \sum_{j=1}^{n} \left(U_t^i U_t^j \left(-X_t^j dL_t^X - Y_t^j dL_t^Y\right)\right)$$

$$+ \frac{1}{2(U_t \cdot V_t)^3} \sum_{j,k=1}^{n} \left(2U_t^i V_t^j V_t^k - \delta_{ij} V_t^k U_t \cdot V_t - \delta_{ik} V_t^j U_t \cdot V_t\right) 4\frac{U_t^j U_t^k}{\|U_t\|^2} dt$$

$$+ \frac{1}{2(U_t \cdot V_t)^3} \sum_{j,k=1}^{n} \left(2U_t^i U_t^j U_t^k\right) 4\left(\delta_{jk} - \frac{U_t^j U_t^k}{\|U_t\|^2}\right) dt$$

$$= \frac{1}{(U_t \cdot V_t)^2} \sum_{j=1}^{n} \left(U_t^i U_t^j + U_t^i V_t^j - \delta_{ij}\, U_t \cdot V_t\right) X_t^j dL_t^X$$

$$+ \frac{1}{(U_t \cdot V_t)^2} \sum_{j=1}^{n} \left(U_t^i U_t^j - U_t^i V_t^j + \delta_{ij}\, U_t \cdot V_t\right) Y_t^j dL_t^Y.$$

Using the fact that $L_t^Y \equiv 0$ on the time interval $[0, \tau \wedge \tau_1)$ (the process Y_t cannot reach the boundary $\partial \mathbb{U}$ before either coupling first with X_t or before the first time when $\|X_t\| = \|Y_t\| = 1$, that is before $0 \in \mathcal{M}_t$), and that L_t^X increases only when $X_t \in \partial \mathbb{U}$, that is only when $\|X_t\| = \|\frac{U_t + V_t}{2}\| = 1$, we obtain:

$$\frac{1}{2}dA_t^i = \frac{1}{(U_t \cdot V_t)^2} \sum_{j=1}^{n} \left(U_t^i U_t^j + U_t^i V_t^j - \delta_{ij} \, U_t \cdot V_t \right) X_t^j \, dL_t^X$$

$$= \frac{U_t^i}{2(U_t \cdot V_t)^2} \sum_{j=1}^{n} \left(U_t^j + V_t^j \right)^2 dL_t^X - \frac{U_t^i + V_t^i}{2U_t \cdot V_t} dL_t^X$$

$$= \frac{U_t^i}{2(U_t \cdot V_t)^2} \| U_t + V_t \|^2 dL_t^X - \frac{U_t^i + V_t^i}{2U_t \cdot V_t} dL_t^X$$

$$= \frac{2U_t^i}{(U_t \cdot V_t)^2} dL_t^X - \frac{U_t^i + V_t^i}{2U_t \cdot V_t} dL_t^X$$

$$= \frac{1}{U_t \cdot V_t} \left(A_t^i - \frac{U_t^i + V_t^i}{2} \right) dL_t^X,$$

thus proving the claim (9.11).

To prove the claim of the lemma, assume by contradiction that there exists a point $R \in \mathbb{U}$ and times $0 < t_1 < t_2 < \tau \wedge \tau_1$ such that the point R and the origin are separated by $\mathcal{M}_{t_1}$, but are not separated by $\mathcal{M}_{t_2}$. Changing the point R, if necessary, we may assume without loss of generality that $R \notin \mathcal{M}_{t_2}$, and using (9.5) and (9.6) we obtain:

$$R \cdot U_{t_2} - \frac{1}{2} U_{t_2} \cdot V_{t_2} < 0 < R \cdot U_{t_1} - \frac{1}{2} U_{t_1} \cdot V_{t_1},$$

or equivalently (recall the definition (9.10) of the process A_t and that $U_t \cdot V_t > 0$ for $t \in [0, \tau \wedge \tau_1)$)

$$R \cdot A_{t_2} < 1 < R \cdot A_{t_1}.$$

Setting $t_0 = \inf \{t > t_1 : R \cdot A_t < 1\} \in (t_1, t_2)$ and using (9.11), we obtain:

$$R \cdot A_{t_0} = R \cdot A_{t_1} + \int_{t_1}^{t_0} R \cdot dA_t$$

$$= R \cdot A_{t_1} + \int_{t_1}^{t_0} \frac{2}{U_t \cdot V_t} \left(R \cdot A_t - \frac{1}{2} R \cdot (U_t + V_t) \right) dL_t^X$$

$$\geq P \cdot A_{t_1} > 1,$$

since $R \cdot A_t \geq 1$ for $t \in [t_1, t_0]$ and

$$\left| \frac{1}{2} R \cdot (U_t + V_t) \right| = |R \cdot X_t| \leq \| R \| \, \| X_t \| \leq 1.$$

By the continuity of the process A_t and the choice of t_0 we must also have $R \cdot A_{t_0} = 1$, contradiction which concludes the proof of the lemma. □

Theorem 9.3. *For any points $x, y \in \overline{U}$ with $\|y\| \leq \|x\|$ and any $z \in \overline{U}$ such that $\|x - z\| \leq \|y - z\|$, we have:*

$$\mathbb{P}^y \left(\|Y_t - z\| < \varepsilon \right) \leq \mathbb{P}^x \left(\|X_t - z\| < \varepsilon \right), \tag{9.12}$$

for any $t \geq 0$ and $\varepsilon \in (0, \min\{\|z\|, 1 - \|z\|\})$, where X_t and Y_t are reflected Brownian motions in U starting at x, respectively y, and $\mathbb{P}^x$, $\mathbb{P}^y$ denote the corresponding probability measures.

Proof. Without loss of generality we may assume that x and y are distinct points.

Let X_t, Y_t be a mirror coupling of reflected Brownian motions in U with starting points $X_0 = x$ and $Y_0 = y$, and let τ be the coupling time and $\tau_1 = \inf\{t > 0 : 0 \in \mathcal{M}_t\}$.

If $\mathcal{M}_t$ separates X_t and z for some $t < \tau \wedge \tau_1$, there exists a point $R \in U$ such that the origin and the point P are separated by $\mathcal{M}_0$, but are not separated by $\mathcal{M}_t$, contradicting Lemma 9.2. It follows that the mirror $\mathcal{M}_t$ does not separate the points X_t and z for all $t < \tau \wedge \tau_1$, and therefore the distance from X_t to z is not greater than the distance from Y_t to z in this case.

Since for $t \geq \tau \wedge \tau_1$, either the processes X_t and Y_t are symmetric with respect to the (fixed) hyperplane $\mathcal{M}_{\tau \wedge \tau_1}$ passing through the origin (for $t \in (\tau \wedge \tau_1, \tau)$), or they have coupled (for $t \in (\tau, \infty)$), it follows that the distance from X_t to z is also not greater than the distance from Y_t to z.

In all cases we obtained that the distance from X_t to z is not greater than the distance from Y_t to z, and the claim follows. □

Theorem 9.4. *For any $x \in U - \{0\}$, $r \in (0, \min\{\|x\|, 1 - \|x\|\})$ and $t > 0$ we have:*

$$\int_{\partial U} p_U (t, x + ru, x) \, d\sigma(u) \leq p_U(t, x + r\tfrac{x}{\|x\|}, x) \leq p_U(t, x + r\tfrac{x}{\|x\|}, x + r\tfrac{x}{\|x\|}), \tag{9.13}$$

where σ is the normalized surface measure on ∂U.

Proof. Using the continuity of the transition density $p_U(t, x, y)$ of reflected Brownian motion in the space variable, it follows $p_U(t, x, y)$ can be written as

$$p_U(t, x, y) = \lim_{\varepsilon \searrow 0} \frac{1}{c_n \varepsilon^n} \int_{\|y-z\|<\varepsilon} p_U(t, x, z) dz = \lim_{\varepsilon \searrow 0} \frac{1}{c_n \varepsilon^n} \mathbb{P}^x \left(\|W_t - y\| < \varepsilon \right), \tag{9.14}$$

where W_t is a reflected Brownian motion in the unit ball $U \subset \mathbb{R}^n$ starting at $W_0 = x$, $\mathbb{P}^x$ denotes the corresponding probability measure and c_n is the volume of the unit ball $U \subset \mathbb{R}^n$.

It is easy to see that, for any $u \in \mathbb{U}$, the hypotheses of Theorem 9.3 are verified if we replace x, y and z respectively by $x + r\frac{x}{\|x\|}$, $x + ru$ and x. From this theorem, and combining with the above representation, we obtain

$$p_{\mathbb{U}}(t, x + ru, x) \leq p_{\mathbb{U}}(t, x + r\frac{x}{\|x\|}, x), \qquad u \in \mathbb{U}. \tag{9.15}$$

Integrating with respect to $u \in \mathbb{U}$ we obtain the left inequality in (9.13).

The second inequality in (9.13) can be proved similarly, replacing x, y and z in Theorem 9.3 respectively by $x + r\frac{x}{\|x\|}$, x and $x + r\frac{x}{\|x\|}$, and using the symmetry of $p_{\mathbb{U}}(t, x, y)$ in x, $y \in \mathbb{U}$. $\qquad\square$

Remark 9.5. The inequality (9.15) can be interpreted as an extremal property of reflected Brownian motion in the unit ball $\mathbb{U}$, as follows:

$$\max_{y \in \mathbb{U} : \|y - x\| = r} p_{\mathbb{U}}(t, x, y) = p_{\mathbb{U}}(t, x, x + r\frac{x}{\|x\|}),$$

that is, among all reflected Brownian motions in the unit ball $\mathbb{U}$ with starting points on the sphere $\{y \in \mathbb{R}^n : \|y - x\| = r\}$, the reflected Brownian motion starting closest to the boundary of $\mathbb{U}$ (i.e. at the point $x + r\frac{x}{\|x\|}$) is most likely to return to (a neighborhood of) x.

Proof of Theorem 9.1. First note that for any fixed $t > 0$, by the radial symmetry of the problem it follows that $p_{\mathbb{U}}(t, x, x)$ is a function of $\|x\| \in [0, 1]$.

For any $x \in \mathbb{U} - \{0\}$, from Theorem 9.4 we obtain

$$p_{\mathbb{U}}(t, x + r\frac{x}{\|x\|}, x + r\frac{x}{\|x\|}) - p_{\mathbb{U}}(t, x, x) \geq \int_{\partial \mathbb{U}} p_{\mathbb{U}}(t, x + ru, x) \, d\sigma(u)$$

$$- p_{\mathbb{U}}(t, x, x)$$

$$= \int_{\partial \mathbb{U}} (p_{\mathbb{U}}(t, x + ru, x)$$

$$- p_{\mathbb{U}}(t, x, x)) \, d\sigma(u),$$

for any $r \in (0, \min\{\|x\|, 1 - \|x\|\})$. Dividing by r and passing to the limit with $r \searrow 0$, we obtain:

$$\frac{d}{d\|x\|} p_{\mathbb{U}}(t, x, x) = \lim_{r \searrow 0} \frac{p_{\mathbb{U}}(t, x + r\frac{x}{\|x\|}, x + r\frac{x}{\|x\|}) - p_{\mathbb{U}}(t, x, x)}{r}$$

$$\geq \lim_{r \searrow 0} \int_{\partial \mathbb{U}} \frac{p_{\mathbb{U}}(t, x + ru, x) - p_{\mathbb{U}}(t, x, x)}{r} \, d\sigma(u).$$

Let $\nabla p_{\mathbb{U}}$ denote the gradient of $p_{\mathbb{U}}(t, \cdot, x)$ in the second variable. The function $p_{\mathbb{U}}(t, \cdot, x)$ is C^2 in the second variable, hence $\nabla p_{\mathbb{U}}(t, \cdot, x)$ is bounded in a neighborhood of x. By the bounded convergence theorem, we obtain

$$\frac{d}{d\,\|x\|} p_{\mathbb{U}}\,(t, x, x) \geq \int_{\partial \mathbb{U}} \nabla p_U\,(t, x, x) \cdot u\,d\sigma(u) = 0.$$

Since $x \in \mathbb{U} - \{0\}$ is arbitrary, we have

$$\frac{d}{d\,\|x\|} p_{\mathbb{U}}(t, x, x) \geq 0, \qquad x \in (0, 1),$$

which shows that $p_{\mathbb{U}}\,(t, x, x)$ is a non-decreasing function of $\|x\| \in (0, 1)$, and by continuity this also holds for $\|x\| \in [0, 1]$.

Since $p_{\mathbb{U}}\,(t, x, x)$ is the diagonal of the heat kernel of an operator with real analytic coefficients, $p_{\mathbb{U}}\,(t, x, x)$ is a real analytic function. If $p_{\mathbb{U}}\,(t, x, x)$ were constant on a non-empty open subset of $\overline{\mathbb{U}}$, then it would be identically constant in $\overline{\mathbb{U}}$, which is impossible (it can be shown that $p_{\mathbb{U}}(t, 0, 0) < p_{\mathbb{U}}(t, 1, 1)$ for any $t > 0$). This, together with the fact that $p_{\mathbb{U}}\,(t, x, x)$ is a non-decreasing radial function shows that $p_{\mathbb{U}}\,(t, x, x)$ is in fact a strictly increasing radial function for any $t > 0$, concluding the proof. □

Chapter 10
Reflected Brownian Motion in Time Dependent Domains

This chapter is based on [BCS03]. We will consider domains of the form $\dot{D} = \{(t,x) : t > 0, x < g(t)\}$ or $\dot{D} = \{(t,x) : t > 0, g_1(t) < x < g_2(t)\}$, where $g(t), g_1(t)$ and $g_2(t)$ are continuous functions, unless stated otherwise. When we refer to "reflected Brownian motion" X_t in $\dot{D}$, we will mean the process constructed from the usual Brownian motion using the Skorohod Lemma (see Lemma 10.12 below). The initial distribution of X_t will be denoted $\mu(dx)$. We will sometimes consider σ-finite rather than probability measures $\mu(dx)$—this poses no technical problems. The density of the distribution of X_t will be denoted $u(t,x)$. If we mention $u(0,x)$, it will mean that we implicitly assume that $\mu(dx)$ has a density and that $\mu(dx) = u(0,x)dx$. Note that in view of [BCS04, Theorem 3.8], $u(t,x)$ solves the heat equation inside $\dot{D}$.

When g is of class C^3, we know from [BCS04, Theorem 2.10 and Remark 2.11] that $u(t,x)$ is the only solution of the heat equation in $\dot{D}$ with a given initial condition $u(0,x)$ and no flux through any part of the boundary. We do not know whether this is true for a domain $\dot{D}$ with a boundary defined by an arbitrary continuous functions. In this section, the only heat equation solution we will consider will be the one defined by the reflected Brownian motion, as indicated above. Many examples are concerned with properties of the heat equation solution at time $t = 1$; in many cases g is smooth up to that time and so the heat equation solution is unique on $[0,1]$.

When the domain contracts rapidly, the heat inside the domain is compressed, especially near the boundary, so one can ask whether singular behavior of the heat equation in some time-dependent domains may occur. We will show that indeed in some domains a positive amount of heat may be found at a boundary point—we will refer to it as a "heat atom." This corresponds to an atom of the distribution of reflected Brownian motion. More precisely, consider a domain $\dot{D} = \{(t,x) : t \geq 0, x < g(t)\}$, where g is a continuous function. By Lemma 10.12, for any given standard Brownian motion B_t and $x_0 \leq g(0)$, there is a unique pair (X_t, L_t) of continuous processes such that

K. Burdzy, *Brownian Motion and its Applications to Mathematical Analysis*,
Lecture Notes in Mathematics 2106, DOI 10.1007/978-3-319-04394-4_10,
© Springer International Publishing Switzerland 2014

$$X_t \leq g(t) \quad \text{and} \quad X_t = x_0 + B_t - L_t \quad \text{for all } t \geq 0,$$

and $t \to L_t$ is a non-decreasing process with $L_0 = 0$ that increases only when $X_t = g(t)$, that is,

$$L_t = \int_0^t 1_{\{X_s = g(s)\}} dL_s.$$

Process X_t is called the reflected Brownian motion in $\dot{D}$ driven by B_t, and process L_t is called the boundary local time of X_t. The law of X_t will be denoted as $\mathbb{P}^{x_0}$, and the probability expectation under it as $\mathbb{E}_{x_0}$. For a σ-finite measure μ on $(-\infty, g(0)]$, we define $\mathbb{P}^\mu = \int_{(-\infty, g(0)]} \mathbb{P}^x \mu(dx)$. We will say that the point $(t, g(t))$ with $t > 0$ is a *heat atom* if $\mathbb{P}(X_t = g(t)) > 0$. If there is no danger of confusion, we will simply say that $g(t)$ is a heat atom. The definition is independent of the initial distribution μ, according to the following lemma.

Lemma 10.1. *For any σ-finite measures μ_1 and μ_2 we have $\mathbb{P}^{\mu_1}(X_t = g(t)) > 0$ if and only if $\mathbb{P}^{\mu_2}(X_t = g(t)) > 0$.*

Our first result on heat atoms requires the notion of an upper class function. A function $f(t)$ belongs to the *upper class* for Brownian motion if $\mathbb{P}(\exists \varepsilon > 0 \; \forall t \in (0, \varepsilon) : B_t < f(t)) = 1$, where B_t is a standard one-dimensional Brownian motion with $B_0 = 0$ [Kni81, p. 144]. The function $f(t)$ has to be defined only on some interval of the form $(0, \varepsilon)$, where $\varepsilon > 0$. The upper class is non-empty—see Remark 10.3 (ii) below.

Theorem 10.2. *Suppose that $t > 0$. The point $(t, g(t))$ is a heat atom if and only if the function $s \to g(t - s) - g(t)$ is an upper function for Brownian motion. Thus, heat atoms exist in some domains.*

Remark 10.3. (i) The paper [STW00] is devoted to a seemingly unrelated subject but the ideas and arguments in our Theorems 10.2 and 10.5 are very close to some results in that paper.

(ii) An explicit analytic criterion due to Kolmogorov (see [Kni81, p. 144]) for upper class functions is known, under some regularity assumptions on f. We recall it for the convenience of the reader. Assume that $t^{-1/2} f(t)$ is decreasing. Then the function $f(t)$ is in the upper class if and only if for any $\varepsilon > 0$ we have

$$\int_0^\varepsilon t^{-3/2} f(t) \exp(-f^2(t)/2t) dt < \infty. \tag{10.1}$$

The celebrated Hinchin's Law of Iterated Logarithm [KS91, p. 112] follows from this criterion, because the function $t \to a\sqrt{2t \log\log(1/t)}$ is in the upper class if and only if $a > 1$. Hence, we have the following result.

Corollary 10.4. *Suppose that $a > 0$, $g(t) = a\sqrt{2(1-t)\log\log(1/(1-t))}$ for $t \in [0, 1)$ and $g(t) = 0$ for $t \geq 1$. Then $g(1)$ is a heat atom if and only if $a > 1$.*

Our next theorem is concerned with a (random) domain with a rough boundary. Note that the Kolmogorov criterion does not help to determine what happens in domains $\dot{D}$ with rough boundaries.

Theorem 10.5. *Suppose that B_t is a standard Brownian motion with $B_0 = 0$ and let $g(t) = B_t$ for $t \geq 0$. Then, with probability 1, there is no $t > 0$ such that $g(t)$ is a heat atom.*

Let $\mathcal{A}_g$ be the set of all $t > 0$ such that $(t, g(t))$ is a heat atom. Recall the definition of the Hausdorff dimension—we say that a set K has Hausdorff dimension $\dim(K) = \alpha$ if α is the infimum of numbers β with the following property. For every $\delta > 0$ there exists a countable covering of K with balls with radii r_j, such that $\sum_j r_j^\beta < \delta$.

Theorem 10.6. *(i) For every continuous function $g(t)$, $\dim(\mathcal{A}_g) \leq 1/2$.*
(ii) There exists a continuous function $g(t)$ with $\dim(\mathcal{A}_g) = 1/2$.

Theorem 10.6 (ii) implies that for some $g(t)$, the cardinality of $\mathcal{A}_g$ is c, i.e., it is the same as the cardinality of the real line. Part (i) of the same theorem shows that the Lebesgue measure of $\mathcal{A}_g$ is zero for every $g(t)$. A standard application of the Fubini Theorem yields the following result.

Corollary 10.7. *The random set $\{t \geq 0 : X_t = g(t)\}$ has zero Lebesgue measure, a.s. That is, almost surely, the reflected Brownian motion in $\dot{D}$ spends zero Lebesgue amount of time on the boundary.*

One can imagine another type of singularity for the heat equation besides a heat atom. Recall that $u(t, x)$ denotes the density of the distribution of X_t. The same rapid contraction which causes the heat atom to appear may also induce $u(t, x)$ to take large values for x strictly inside $D(t)$. We will say that $u(t, x)$ is *singular at* $g(t)$ if

$$\limsup_{x \uparrow g(t)} u(t, x) = \infty.$$

The existence of a singularity is the property of the domain and does not depend on the initial condition, as shown in the next result.

Lemma 10.8. *Suppose that μ_1 and μ_2 are σ-finite measures such that $\mu_k([g(0) - a, g(0)]) \leq c_1 a$ for some constant $c_1 < \infty$ and all $a \geq 1$. Let $u_1(t, x)$ and $u_2(t, x)$ be the heat equation solutions corresponding to the initial conditions μ_1 and μ_2.*

(i) For every fixed $t > 0$, $\limsup_{x \uparrow g(t)} u_1(t, x) = \infty$ if and only if $\limsup_{x \uparrow g(t)} u_2(t, x) = \infty$.
(ii) For every fixed $t > 0$, $\lim_{x \uparrow g(t)} u_1(t, x) = \infty$ if and only if $\lim_{x \uparrow g(t)} u_2(t, x) = \infty$, in the sense that if one of the limits exists and is infinite then the same is true of the other.
(iii) If $g(s)$ is non-increasing on the interval $[0, t]$ then $\lim_{x \uparrow g(t)} u_1(t, x) = \infty$ if and only if $\limsup_{x \uparrow g(t)} u_1(t, x) = \infty$.

For the proof see [BCS03].

Intuition suggests that there might be a logical relationship between the existence of a singularity of $u(t, x)$ at $g(t)$ and a heat atom at $g(t)$. Somewhat surprisingly, our next theorem shows that there is no such relationship.

Theorem 10.9. *There exist domains* $\dot{D}_k = \{(t, x) : t \geq 0, x < g_k(t)\}, k = 1, 2, 3, 4$, *such that* $g'_k(1) = -\infty$, *and with the following properties. For definiteness, let* $u_k(t, x)$ *be the heat equation solution in* $\dot{D}_k$ *corresponding to* $u_k(0, x) \equiv 1$.

  (i) $u_1(1, x)$ *is singular at* $g_1(1)$ *and* $g_1(1)$ *is a heat atom.*
  (ii) $u_2(1, x)$ *is singular at* $g_2(1)$ *but* $g_2(1)$ *is not a heat atom.*
  (iii) $u_3(1, x)$ *is not singular at* $g_3(1)$ *but* $g_3(1)$ *is a heat atom.*
  (iv) $u_4(1, x)$ *is not singular at* $g_4(1)$ *and* $g_4(1)$ *is not a heat atom.*

The next result provides some examples of functions satisfying Theorem 10.9 (ii)–(iv).

Theorem 10.10. *Let* $h(t) = \sqrt{t} |\log t|^{\beta}$. *Fix any integer* $k_0 = k_0(\beta) > 1$ *such that* $h'(t) > 0$ *for* $t \leq 2^{-k_0+1}$. *Suppose that* $\dot{D} = \{(t, x) : t \geq 0, x < g(t)\}$ *where* $g(1 - t) = h(t)$ *for* $t \in [0, 2^{-k_0}]$ *and* $g(t) = g(1 - 2^{-k_0})$ *for* $t \in [0, 1 - 2^{-k_0}]$.

  (i) *If* $\beta \in (-\infty, -1)$ *then* $g(1)$ *is not a heat atom and* $u(1, x)$ *is not singular at* $g(1)$.
  (ii) *If* $\beta \in [-1, 0)$ *then* $g(1)$ *is not a heat atom but* $u(1, x)$ *is singular at* $g(1)$.
    Now suppose that $g(t)$ *is a piecewise linear function with vertices at the points* $(1 - 2^{-k}, h(2^{-k}))$ *for* $k \geq k_0$, *and* $(0, h(2^{-k_0}))$.
  (iii) *If* $\beta \in (1/2, \infty)$ *then* $g(1)$ *is a heat atom but* $u(1, x)$ *is not singular at* $g(1)$.

We believe that part (iii) of Theorem 10.10 holds for the function $g(1 - t) = \sqrt{t} |\log t|^{\beta}$, where $\beta > 1/2$, because the ratio of the this function and the function $g(1 - t)$ defined in part (iii) is bounded. However, "sandwiching" a smooth function between two piecewise linear functions for which (iii) holds is not sufficient—there is no obvious domain monotonicity when it comes to the existence of singularities of $u(t, x)$. In view of this lack of domain monotonicity, the following result is considerably stronger than Theorem 10.10 (i).

Proposition 10.11. *Suppose that* g *is a continuous function on* $\mathbb{R}_+$ *and that* $\dot{D} = \{(t, x) : t \geq 0, x < g(t)\}$. *If* $\int_0^1 |g'(s)|^2 < \infty$ *and* $\int_0^1 |g'(1 - s)| s^{-1/2} ds < \infty$ *then there is no heat atom nor heat singularity at* $g(1)$.

It would be natural to conjecture that, in the setting of Theorem 10.10, the domain $\dot{D}$ has a heat atom and singularity at $g(1)$ if $\beta \in (0, 1/2)$. However, heuristic estimates suggest that there is no singularity in this case. The challenge of finding an explicit $g(t)$ such that there is a heat atom at $(1, g(1))$ and $u(t, x)$ is singular at the same point is left as an open problem.

Theorem 10.2 provides, in a sense, a full characterization of heat atoms, especially when it is combined with explicit results contained in (10.1) and Corollary 10.4. Equally explicit characterization of the points where $u(t, x)$ is singular seems to be a much harder problem. One reason why it might be very hard is that there is no monotonicity result for domains with singularities. Let us try to elucidate this point. If we have domains $\dot{D}_k = \{(t, x) : t > 0, x < g_k(t)\}$, $k = 1, 2$, such that $g_1(1) = g_2(1)$ and $g_1(t) \geq g_2(t)$ for $t < 1$, and $\dot{D}_2$ has a heat atom at $(1, g_2(1))$ then so does $\dot{D}_1$. Examples given in Theorem 10.10 show that a similar statement is false for singularities—whether $j = 1$ or 2, the fact that $\dot{D}_j$ has a singularity at $(1, g_j(1))$ does not imply that $\dot{D}_{3-j}$ has a singularity at $(1, g_{3-j}(1))$.

Proofs of some of the above results are given in Sect. 10.2.

10.1 General Results on Reflected Brownian Motion in Time Dependent Domains

This subsection collects some results from [BCS04]. The following lemma is a generalization of the Skorohod decomposition, see [KS91, Lemma 3.6.14]. Let $v^+ = \max\{v, 0\}$ and $v^- = \max\{-v, 0\}$.

Lemma 10.12. *Suppose that g is a locally bounded measurable function from $\mathbb{R}_+$ to $\mathbb{R}$. Let $\hat{g}(t) = \max\left(g(t), \limsup_{s\downarrow t} g(s)\right)$. For every continuous function $b(t)$, $t \geq 0$, there is a unique pair of functions $(x(t), l(t))$, $t \geq 0$, such that*

(i) $x(t) := b(t) + l(t) \geq \hat{g}(t)$ for $t \geq 0$,
(ii) $l(t)$ is a nondecreasing right-continuous function with $l(0) = (b(0) - \hat{g}(0))^-$,
(iii) if $x(t) > \hat{g}(t)$ for $t \in [s_1, s_2]$ then $l(s_1) = l(s_2)$, and
(iv) if $l(t)$ has a jump at $t = t_1$, i.e., $\lim_{t\uparrow t_1} l(t) < l(t_1)$ then $x(t_1) = \hat{g}(t_1)$.

Moreover,

$$l(t) = \sup_{0 \leq s \leq t} (b(s) - \hat{g}(s))^-. \tag{10.2}$$

Proof. We first prove uniqueness. Suppose that $(x(t), l(t))$ and $(\tilde{x}(t), \tilde{l}(t))$ have properties (i)–(iv) and that for some t_* we have $l(t_*) - \tilde{l}(t_*) > 0$. By the right-continuity of l and $\tilde{l}$, there exist $t_1 > t_*$ and $a > 0$ such that $l(t_1) - \tilde{l}(t_1) = a$ and $l(t) - \tilde{l}(t) > 0$ for all $t \in [t_*, t_1]$. Let $t_0 = \inf\{t \leq t_1 : x(s) > \tilde{x}(s) \ \forall s \in [t, t_1]\}$. For every $t \in (t_0, t_1]$ we have $x(t) > \tilde{x}(t) \geq \hat{g}(t)$ so (iii) implies that $l(t) = l(t_1)$. This implies that, for $t \in (t_0, t_1]$,

$$0 < a = x(t_1) - \tilde{x}(t_1) = l(t_1) - \tilde{l}(t_1) \leq l(t) - \tilde{l}(t) = x(t) - \tilde{x}(t). \tag{10.3}$$

Since b, l and $\tilde{l}$ are right-continuous, so are x and $\tilde{x}$ and so (10.3) in fact holds for all $t \in [t_0, t_1]$. In particular, $x(t_0) - \tilde{x}(t_0) \geq a > 0$, which implies, in view of (ii), that $t_0 > 0$. The definition of t_0 and the fact that $x(t_0) - \tilde{x}(t_0) \geq a > 0$ require that $\liminf_{t\uparrow t_0} x(t) - \tilde{x}(t) \leq 0$. Another consequence of (10.3) is that $l(t_0) - \tilde{l}(t_0) \geq a$. We obtain

$$\left(\liminf_{t \uparrow t_0} l(t)\right) - (l(t_0) - a) \le \left(\liminf_{t \uparrow t_0} l(t)\right) - \tilde{l}(t_0) \le \liminf_{t \uparrow t_0}(l(t) - \tilde{l}(t))$$

$$= \liminf_{t \uparrow t_0}(x(t) - \tilde{x}(t)) \le 0.$$

We see that $\lim_{t \uparrow t_0} l(t) < l(t_0)$ and so, according to (iv), $x(t_0) = \hat{g}(t_0)$. However, $\tilde{x}(t_0) \le x(t_0) - a = \hat{g}(t_0) - a$. This implies that $\tilde{x}(t) < \hat{g}(t)$ for some t, which is a contradiction. The proof of uniqueness is complete.

We will finish the proof by showing that $l(t)$ defined in (10.2), together with $x(t) = b(t) + l(t)$, satisfy (i)–(iv). Property (i) is evident and so is the fact that $l(t)$ is non-decreasing. It is easy to see that $\hat{g}(t) \ge \limsup_{s \downarrow t} \hat{g}(s)$. Right-continuity of $l(t)$ easily follows from this observation.

To prove (iv), note that by the continuity of $b(t)$ and right-continuity of $l(t)$, $x(t)$ is right-continuous and so $\lim_{t \downarrow t_1} x(t)$ exists and is equal to $x(t_1)$. Let

$$t_* = \sup\{s : \sup_{0 \le r \le s} (b(s) - \hat{g}(s))^- = 0\}.$$

Then $(x(t), l(t)) = (b(t), 0)$ is the unique solution to the Skorohod problem on $[0, t_*)$ and so $l(t)$ has no jumps on this interval. Suppose now that $t_1 \ge t_*$ is a jump time for l. Let $b(t_1) = c$. Then $l(t_1) = -c + \hat{g}(t_1)$ because of continuity of $b(t)$ and the fact that $l(t)$ has a jump at $t = t_1$. We conclude that

$$x(t_1) = b(t_1) + l(t_1) = c - c + \hat{g}(t) = \hat{g}(t),$$

which proves (iv).

It remains to prove (iii). Suppose that $x(t) > \hat{g}(t)$ for $t \in [s_1, s_2]$. Then $l(t) > \hat{g}(t) - b(t)$ on the same interval. Since

$$l(s_2) = \max\left\{\sup_{0 \le s \le s_1} (b(s) - \hat{g}(s))^-, \sup_{s_1 \le s \le s_2} (b(s) - \hat{g}(s))^-\right\},$$

we must have $l(s_2) = l(s_1)$.   □

Corollary 10.13. *In the setting of Lemma 10.12, consider the Skorohod problem for a fixed function $b(t)$, relative to two different measurable functions $g_1(t)$ and $g_2(t)$. Let $(x_1(t), l_1(t))$ and $(x_2(t), l_2(t))$ denote the corresponding solutions of the Skorohod problem. If $g_1(t) \le g_2(t)$ for all t then $x_1(t) \le x_2(t)$ for all t. If $|g_1(t) - g_2(t)| \le \varepsilon$ for all t then $|x_1(t) - x_2(t)| \le \varepsilon$ for all t.*

Corollary 10.14. *Suppose that we have a family of continuous functions $g_\alpha(t)$, where α is an index in some metric space and assume that the mapping $\alpha \to g_\alpha(\cdot)$ is continuous in the uniform topology. Let $(x_\alpha(t), l_\alpha(t))$ denote solutions of the Skorohod problem for a fixed $b(t)$ (same for all α), relative to $g_\alpha(t)$. Then the mapping $\alpha \to x_\alpha(t)$ is continuous in the uniform topology.*

The two corollaries stated above follow easily from formula (10.2).

Theorem 10.15. *Suppose that g is a continuous function on $\mathbb{R}_+$ and that $\dot{D} = \{(t, y) : t \geq 0, y < g(t)\}$. Let X be the reflected Brownian motion in $\dot{D}$ with initial distribution X_0 being the Lebesgue measure on $(-\infty, g(0))$. Let B_t be the standard Brownian motion and $Y_t = Y_0 + B_t - L_t$ be the reflected Brownian on $(-\infty, 0]$, with L_t the local time of Y at 0. Let $\mathbb{P}^x$ denote the law of Y with $Y_0 = x$. Assume that $\int_0^1 |g'(s)|^2 ds < \infty$ and let*

$$N_t = \exp\left(\int_0^t g'(t - s)dB_s - \frac{1}{2}\int_0^t |g'(t-s)|^2 ds - 2\int_0^t g'(t-s)dL_s\right), \quad 0 \leq t \leq 1.$$

$$(10.4)$$

Then for each $t \in [0, 1]$, $\mathbb{E}_x[N_t]$ is bounded on compact intervals of $(-\infty, 0)$ and the distribution of X_t is absolutely continuous with respect to the Lebesgue measure on $(-\infty, g(t)]$ with density function u given by $u(t, g(t) + x) = \mathbb{E}_x[N_t]$.

For the proof, see [BCS04].

10.2 Proofs of Theorems on Heat Atoms and Singularities

Proof of Lemma 10.1 and Theorem 10.2. Fix some $t > 0$ and consider a Brownian motion B_r with the initial distribution μ. First suppose that the function $s \to g(t - s) - g(t)$ is not an upper function for Brownian motion B_r. By the definition of the upper function and Brownian invariance under time reversal, with probability 1,

$$M := \sup_{s \in (0,t)} B(t - s) - B(t) - (g(t - s) - g(t)) > 0.$$

By the continuity of $s \to B(t-s) - B(t) - (g(t-s) - g(t))$, there exists $s_0 \in (0, t]$ such that

$$B(t - s_0) - B(t) - (g(t - s_0) - g(t)) = M,$$

and for $s \in [0, s_0)$, we have

$$B(t - s) - B(t) - (g(t - s) - g(t)) < M.$$

Hence,

$$B(t - s) - g(t - s) < B(t - s_0) - g(t - s_0),$$

for $s \in [0, s_0)$. In other words,

$$B(r) - B(t - s_0) < g(r) - g(t - s_0), \quad (10.5)$$

for $r \in (t - s_0, t]$.

We can suppose that the reflected Brownian motion X_r in $\dot{D}$ is constructed from B_r by the means of Lemma 10.12. Hence, $X_r = B_r + L_r$, for $r > 0$, where L_r is non-increasing and does not decrease when $X_r < g(r)$. Let $\hat{X}_r = X_r$ for $r \in [0, t - s_0]$ and $\hat{X}_r = X_{t-s_0} + B_r - B_{t-s_0}$ for $r \in (t - s_0, t]$. We let $\hat{L}_r = L_r$ for $r \in [0, t - s_0]$ and $\hat{L}_r = L_{t-s_0}$ for $r \in (t - s_0, t]$. In view of (10.5) and the fact that $X_{t-s_0} \le g(t - s_0)$, we have for $r \in (t - s_0, t]$,

$$\hat{X}_r = X_{t-s_0} + B_r - B_{t-s_0} < X_{t-s_0} + g(r) - g(t - s_0) \le g(r).$$

This implies that $(\hat{X}_r, \hat{L}_r)$ solves the Skorohod problem for $r \in (t - s_0, t]$. By the uniqueness of the solution of the Skorohod problem, $\hat{X}_r = X_r$ for $r \in (t - s_0, t]$. Hence, $X_r < g(r)$ for $r \in (t - s_0, t]$. In particular, $X_t < g(t)$, a.s. This completes the proof of the "only if" part of Theorem 10.2. In relation to Lemma 10.1, note that the above argument applies to all initial distributions μ.

We will now assume that $s \to g(t - s) - g(t)$ is an upper function for Brownian motion B_r. We use time reversal as in the first part of the proof to see that with a strictly positive probability, the process $s \to B(t - s) - B(t)$ stays below the graph of $s \to g(t - s) - g(t)$ for all $s \in (0, t]$.

Suppose that $\mu((-\infty, g(0)) > 0$. The case when μ is concentrated at $g(0)$ can be easily handled using the Markov property and the fact that at any time $s > 0$, $u(s, x) > 0$ for $x < g(s)$. Recall that B_r is a Brownian motion with the initial distribution μ. Let A be the event that $B(t) > g(t)$ and $B(t - s) - B(t) < g(t - s) - g(t)$ for all $s \in (0, t)$. In view of the assumption that $\mu((-\infty, g(0)) > 0$, it is easy to see that A has a positive probability. Recall the Skorohod representation $X_r = B_r + L_r$, as in the first part of the proof, and suppose that the event A occurred. We will argue that we must have $X_t = g(t)$. Suppose it is not so. Since A occurred, we have $B(t) > g(t)$, so there exists an s with $X_s = g(s)$. Let $t_0 := \sup\{s < t : X_s = g(s)\}$ and note that $t_0 < t$, because we assumed that $X_t < g(t)$. It follows from the definition of t_0 that $X_r < g(r)$ for $r \in (t_0, t]$. This implies that $L_r = L_{t_0}$ for all $r \in (t_0, t]$. In view of this observation and our assumption that $X_t < g(t)$,

$$B(t_0) - B(t) = X_{t_0} - X_t > g(t_0) - g(t).$$

But this contradicts the definition of A. Hence, $\mathbb{P}(X_t = g(t)) \ge \mathbb{P}(A) > 0$. This proves the "if" part of Theorem 10.2 and also finishes the proof of Lemma 10.1, since the argument does not depend on μ. $\square$

Proof of Theorem 10.5. We will show that if $f(t)$ is an upper function for Brownian motion then $f(t) > 0$ for t in some interval $(0, \delta)$ where $\delta > 0$. Suppose that there is a sequence $t_n \downarrow 0$, such that $f(t_n) \le 0$. Let A_n be the event $\{B(t_n) \ge f(t_n)\}$. Clearly, $\mathbb{P}(A_n) \ge 1/2$. This implies that for every k, $\mathbb{P}(\exists t \in (0, 1/k) : B_t \ge f(t)) \ge 1/2$. Hence, the event $\bigcap_{k \ge 1}\{\exists t \in (0, 1/k) : B_t \ge f(t)\}$ has probability greater than or equal to $1/2$. By the 0-1 law, its probability is equal to 1, which shows that $f(t)$ is not in the upper class.

We see that the only points $g(t) = B_t$ which might be heat atoms are left sided local minima, i.e., the time t must have the property that $B_s > B_t$, for all $s \in (t-\varepsilon, t)$ and some $\varepsilon > 0$. Let $\hat{B}_t$ be a Brownian motion independent of B_t. Suppose $t_0 > 0$ is such that $g(t_0)$ is a heat atom, i.e., $s \to B(t_0 - s) - B(t_0)$ is an upper function. By the definition of an upper function, a.s.,

$$\hat{B}(t_0 - s) - \hat{B}(t_0) < B(t_0 - s) - B(t_0), \qquad (10.6)$$

for s in a small interval $(0, \varepsilon_1)$. Since $-\hat{B}_t$ is also a Brownian motion, a.s.,

$$(-\hat{B}(t_0 - s)) - (-\hat{B}(t_0)) < B(t_0 - s) - B(t_0), \qquad (10.7)$$

for $s \in (0, \varepsilon_2)$. The fact that $B(t_0)$ is a left sided local minimum and conditions (10.6) and (10.7) imply that the trajectory of the two-dimensional process $(B_s, \hat{B}_s)$ stays strictly inside a cone with vertex $(B_{t_0}, \hat{B}_{t_0})$ and right angle, for $s \in (t_0 - \varepsilon_3, t_0)$. However, Brownian motion has no "cone points" with the right angle a.s. (see [Shi85]). This proves Theorem 10.5. □

Proof of Theorem 10.6. (i) We will first analyze monotone functions $g(t)$. Our first step is to prove a statement slightly stronger than that at the beginning of the proof of Theorem 10.5.

Fix any $\alpha > 1/2$. We will show that if $f(t)$ is an upper function for Brownian motion then $f(t) > t^\alpha$ for t in some interval $(0, \delta)$ where $\delta > 0$. Suppose that there is a sequence $t_n \downarrow 0$, such that $f(t_n) \le t_n^\alpha$. Let A_n be the event $\{B(t_n) \ge f(t_n)\}$. Recall that $B(t_n)$ is a normal random variable with the standard deviation equal to $t_n^{1/2}$. Hence, for large n, $\mathbb{P}(A_n) \ge 1/4$. This implies that for every k, $\mathbb{P}(\exists t \in (0, 1/k) : B_t \ge f(t)) \ge 1/4$. Therefore, the event $\bigcap_{k \ge 1} \{\exists t \in (0, 1/k) : B_t \ge f(t)\}$ must have probability greater than or equal to $1/4$. By the 0-1 law, its probability is equal to 1, which shows that $f(t)$ is not in the upper class.

Let $\mathcal{A}^j = \mathcal{A}_g \cap [0, j]$. In order to prove that $\dim(\mathcal{A}_g) \le 1/2$, it will suffice to show that $\dim(\mathcal{A}^j) \le 1/2$ for every j. We will consider only $\mathcal{A}^1$; the other sets can be treated in a completely analogous way.

Recall that we decided to focus our attention on monotone functions $g(t)$ first. Of course, increasing functions $g(t)$ have no heat atoms, so let us assume that $g(t)$ is non-increasing. Fix any $\alpha \in (1/2, 1)$ and $\delta > 0$. For every heat atom $(t, g(t))$ with $t \in (0, 1)$, let I_t be the largest subinterval of $(0 \vee (t - \delta), t)$ of the form (s, t), such that $g(t - v) - g(t) > v^\alpha$ for all $t - v \in (s, t)$. The interval I_t is non-degenerate, by the introductory remarks in this proof. Let I be the union of all intervals I_t. Since I is an open set, it is a disjoint union of open intervals J_k. Consider any closed interval $\tilde{J}_k \subset J_k$. Since the intervals I_t cover $\tilde{J}_k$ and $\tilde{J}_k$ is compact, there is a finite subcovering of $\tilde{J}_k$ consisting of, say, $I_{t_1}, I_{t_2}, \ldots, I_{t_m}$. Let $J_k^* = I_{t_1} \cup I_{t_2} \cup \cdots \cup I_{t_m}$ and suppose that $I_{t_j} = (a_j, b_j)$. We will suppose without loss of generality that $b_{j+1} \le b_j$ for all j. Then we can eliminate redundant I_{t_j}'s from the covering of $\tilde{J}_k$ so that $a_{j+1} \leq a_j$ for all j. It is easy to see that we must have $b_{j+1} > a_j$ for all j. By the definition of I_{t_j}, $g(b_{j+1}) - g(b_j) > (b_j - b_{j+1})^\alpha$. Since this holds for all $j < m$ and we also have $g(a_m) - g(b_m) \geq (b_m - a_m)^\alpha$, it is not hard to see that $g(a_m) - g(b_1) \geq (b_1 - a_m)^\alpha$. Let us denote the endpoints of J_k by $c_k - r_k$ and c_k, with $r_k > 0$. Since $\tilde{J}_k$ is an arbitrarily large closed subinterval of J_k and $\tilde{J}_k \subset J_k^* \subset J_k$, we see, using the continuity of $g(t)$, that $g(c_k - r_k) - g(c_k) \geq r_k^\alpha$. The function $g(t)$ is monotone and the intervals J_k are disjoint so $\sum_k g(c_k - r_k) - g(c_k) \leq g(0) - g(1)$. We conclude that

$$\sum_k r_k^\alpha \leq g(0) - g(1). \tag{10.8}$$

Now recall that the length of I_{t_j} does not exceed δ. This implies that the line passing through $(b_{j+1}, g(b_{j+1}))$ and $(b_j, g(b_j))$ must have a slope smaller than $-\delta^\alpha/\delta = -\delta^{\alpha-1}$. By linking consecutive points of the form $(b_j, g(b_j))$ we obtain an estimate for the slope of the line passing through $(a_m, g(a_m))$ and $(b_1, g(b_1))$—it is also bounded above by $-\delta^{\alpha-1}$. However, the decrement of $g(t)$ on any subinterval of $(0, 1)$ is bounded by $g(0) - g(1)$. By approximating J_k by large subintervals $\tilde{J}_k$, we see that the length r_k of J_k is bounded by $(g(0) - g(1))\delta^{1-\alpha}$. Hence, for any $\varepsilon > 0$ we can find $\delta > 0$ such that $r_k < \varepsilon$ for every k.

Consider arbitrarily small $\alpha_1 > \alpha$. Assuming that $\delta > 0$ is sufficiently small so that $r_k < \varepsilon$ for all k, we obtain from (10.8),

$$\sum_k r_k^{\alpha_1} \leq \varepsilon^{\alpha_1 - \alpha}(g(0) - g(1)).$$

Note that if $(t, g(t))$ is a heat atom then t must lie inside one of the intervals J_k or be the right endpoint of one of those intervals. Let $\hat{J}_k$ be an interval with the same center as J_k but with twice greater length, $\hat{r}_k = 2r_k$. The intervals $\hat{J}_k$ cover the set $\mathcal{A}^1$ and their radii $\hat{r}_k$ satisfy

$$\sum_k (\hat{r}_k)^{\alpha_1} \leq 2^{\alpha_1} \varepsilon^{\alpha_1 - \alpha}(g(0) - g(1)).$$

The right hand side can be made arbitrarily small by choosing a suitable $\varepsilon > 0$ (and the corresponding $\delta > 0$). We conclude that the Hausdorff dimension of $\mathcal{A}^1$ is less than or equal to α_1. However, α_1 can be chosen arbitrarily close to α and α can be any number greater than $1/2$, so $\dim \mathcal{A}^1 \leq 1/2$.

We have proved part (i) of the theorem for monotone functions $g(t)$. We will now show how to derive the general statement from this preliminary result.

For a continuous function $f : [0, a] \to \mathbb{R}$, we define its future minimum function $(f_a(t), t \in [0, a])$ by $f_a(t) = \min\{f(s) : s \in [t, a]\}$. Assume that $f(0) = f_a(0) = 0$ and $f_a(t) > 0$ for some t. We will show that $f(t)$ is an upper class function if and only if $f_a(t)$ is an upper class function for some $a > 0$. The "if" assertion is clear because $f(t) \geq f_a(t)$. Now suppose that $f_a(t)$ does not belong to the upper class for any $a > 0$. We will show that $f(t)$ does not belong to the upper class either. If for every a there is $b > 0$ such that $f_a(t) \leq 0$ for every $t \in [0, b]$ then $f(t) \leq 0$ for some t in every neighborhood of 0. Such a function cannot be an upper class function, as shown at the beginning of the proof of Theorem 10.5. Assume that for some $a > 0$ we have $f_a(t) > 0$ for all $t > 0$. If $f_a(t) = f(t)$ for all $t \in [0, \varepsilon]$ where $\varepsilon > 0$ then we are done. So suppose that $f_a(t) \neq f(t)$ for some $t > 0$ in every neighborhood of 0. Since $f(t)$ and $f_a(t)$ are continuous, the set of all t such that $f_a(t) < f(t)$ is a disjoint union of open intervals M_k. By continuity, the value of $f_a(t)$ must converge to 0 as $t \to 0$, but it is not equal to 0 on any interval M_k. This shows that there must be infinitely many different intervals M_k in every neighborhood of 0. On every such interval, the function $f_a(t)$ is constant. Suppose that $M_k = (s_0, s_1)$, $f_a(s) = x$ for $s \in (s_0, s_1)$, and consider a Brownian motion starting from $B_s = x$, where $s \in (s_0, s_1)$. This Brownian motion has at least $1/2$ chance of crossing the function $f(t)$ before s_1, because there is $1/2$ chance that it will be above x at time s_1. Consider an arbitrarily small $\varepsilon > 0$ and find an interval M_k contained in $(0, \varepsilon)$. Let ε_1 denote its right endpoint. Find even smaller $\varepsilon_2 > 0$ such that the Brownian motion starting from $B_0 = 0$ will hit $f_a(t)$ between ε_2 and ε_1 with probability greater than $1/2$. By the strong Markov property, the Brownian motion will hit $f(t)$ before ε_1 with probability not less than $1/4$. Since $\varepsilon > 0$ is arbitrarily small, the standard 0-1 argument shows that the Brownian motion must hit $f(t)$ infinitely often in every neighborhood of 0, a.s. This completes the proof of the claim that $f(t)$ is an upper class function if and only if $f_a(t)$ is an upper class function for some $a > 0$.

Now consider an arbitrary continuous $g(t)$. For an interval $[a, b]$ with rational endpoints, let $g_{a,b}(t) = \min\{g(s) : s \in [a, t]\}$. Consider a t_0 such that $(t_0, g(t_0))$ is a heat atom for $g(t)$. Then $g(t) > g(t_0)$ for all $t \in [a, t_0)$, and some rational $a < t_0$. Suppose $b > t_0$ is also a rational number. We see that $g_{a,b}(t) > g(t_0)$ for $t \in [a, t_0)$. Hence, according to the last paragraph, t_0 is the location of a heat atom in the domain $\{(t, x) : t \in [a, b], x < g_{a,b}(t)\}$. However, $g_{a,b}(t)$ is a monotone function and so the family $\mathcal{A}_{a,b}$ of its heat atoms is a set with Hausdorff dimension less than or equal to $1/2$. We have proved that $\mathcal{A}_g \subset \bigcup_{a,b} \mathcal{A}_{a,b}$, where the union is taken over all rational a and b. A countable union of sets with Hausdorff dimension less than or equal to $1/2$ has a Hausdorff dimension not exceeding $1/2$—this completes the proof of part (i).

(ii) First we will prove that for any fixed $\alpha < 1/2$ there exists a (random) function $g(t)$ with $\dim \mathcal{A}_g \geq \alpha$. Suppose that S_t is a stable subordinator with index α (see [Ber96]) starting from $S_0 = 0$ and let $S_t^{-1} = \inf\{u : S_u \geq t\}$. Let $g(t) = S^{-1}(S_1 - t) - 1$ for $t \in [0, S_1]$ and $g(t) = -1$ for $t > S_1$. For $a \in (-1, 0)$, let $g^{-1}(a) = \inf\{t : g(t) \leq a\}$.

Recall that $\alpha < 1/2$ and fix some $\beta \in (2, 1/\alpha)$. For a fixed $t \geq 0$, a.s.,

$$\limsup_{s \downarrow 0} \frac{S_{t+s} - S_t}{s^{\beta}} = 0,$$

according to Theorem 9 and the remark following it in [Ber96]. Hence, for any fixed $a \in (-1, 0)$ and $t_0 = g^{-1}(a)$,

$$g(t_0 - s) - g(t_0) \geq s^{1/\beta},$$

for sufficiently small $s > 0$. Since $1/\beta < 1/2$, Kolmogorov's test (10.1) (see also Corollary 10.4) implies that $(t_0, g(t_0))$ is a heat atom a.s. By Fubini's theorem, with probability 1, for almost all $a \in (0, 1)$ and $t_0 = g^{-1}(a)$, the point $(t_0, g(t_0))$ is a heat atom. Theorem III.16 of [Ber96] implies that the Hausdorff dimension of the set of such t_0 is the same as the Hausdorff dimension of the range of the stable subordinator S_t, i.e., it is equal to α. Thus, $\dim \mathcal{A}_g \geq \alpha$.

A standard splicing argument now shows that there exists a function $g(t)$ such that the set of its heat atoms $t \in (1 - 2^{-n}, 1 - 2^{-n-1})$ has Hausdorff dimension greater than $(1/2) - (1/n)$. It follows that the set of all heat atoms of g in $[0, 1]$ has Hausdorff dimension greater or equal to $1/2$. The dimension must be equal to $1/2$, by part (i) of the theorem.   □

Proof of Theorem 10.9. Existence of functions g_2, g_3 and g_4 satisfying (ii)–(iv) will be established in the proof of Theorem 10.10 below. We also note a trivial example $g_4(t) \equiv 0$.

We will now prove parts (i) and (ii) of the theorem. As we said, part (ii) will be demonstrated in the proof of Theorem 10.10 below but the same argument generates examples in parts (i) and (ii) of Theorem 10.9. According to Theorem 10.2, we can find a function $h_0(t)$ such that $h_0(1)$ is a heat atom in the domain $\dot{D}_0^* = \{(t, x) : t \geq 0, x < h_0(t)\}$. By Corollary 10.4, we can assume that $h_0(t)$ is decreasing and concave, and by shift invariance of the problem we may suppose that $h_0(1) = 0$. Let X_t be the reflected Brownian motion in D_0^* with the initial distribution $\mu(dx) = dx$, the Lebesgue measure on $(-\infty, h_0(0))$. Let q_0 be the size of the heat atom at $(1, 0)$, i.e., $\mathbb{P}(X_1 = 0) = q_0 > 0$.

We will assume that all reflected Brownian motions in the rest of the proof are constructed from the same Brownian motion B_t with the initial distribution μ, by the means of Lemma 10.12.

Fix any $q_\infty \in [0, q_0)$. Let $q_k = q_\infty + (q_0 - q_\infty)/2^k$ for $k \geq 1$. We will construct a sequence of domains $\dot{D}_k^* = \{(t, x) : t \geq 0, x < h_k(t)\}$, for suitably chosen $h_k(t)$. The reflected Brownian motions in $\dot{D}_k^*$ will be denoted X_t^k. We will also choose numbers $d_k \leq (q_{k-1} - q_k)/2^k$. For a $c_k > 0$, let p_k be the probability that the standard Brownian motion B_t differs at time $t = c_k$ from its starting point by more than d_k. For a given d_k we let $c_k > 0$ be the real number such that $q_{k-1} - p_k - q_k = (q_{k-1} - q_k)/2$.

Let $d_1 = (q_0 - q_1)/2$. We will consider a family of domains $\dot{D}_{1,y}^* = \{(t,x) : 0 \le t \le 1, x < h_{1,y}(t)\}$, where $h_{1,y}(t)$ is the minimum of two functions: $h_0(t)$ and the linear function passing through the points $(1 - c_1, h_0(1 - c_1))$ and $(1, y)$. We restrict the values of y in the definition of $\dot{D}_{1,y}^*$ to the interval $[0, h_0(1 - c_1)]$. Note that the boundary of $\dot{D}_{1,0}^*$ is linear on the interval $(1 - c_1, 1)$, so the point $(1, 0)$ is not a heat atom in $\dot{D}_{1,0}^*$, according to Theorem 10.2 and Corollary 10.4. In view of the assumption that reflected Brownian motions $X_t^{*,1,y}$ in domains $\dot{D}_{1,y}^*$ are constructed from the same standard Brownian motion using the Skorohod Lemma and Corollary 10.14, the processes $X_t^{*,1,y}$ converge uniformly a.s. when y's converge to a limit. By the continuity of probability, we can find $y_1 \in (0, h_0(1-c_1))$ such that $\mathbb{P}(X_1^{*,1,y_1} = 0) = q_1$. We let $h_1(t) = h_{1,y_1}(t)$.

We proceed by induction. Suppose that the functions $h_1, h_2, \ldots, h_k$ and the corresponding domains $\dot{D}_1^*, \dot{D}_2^*, \ldots, \dot{D}_k^*$ have been defined. Recall that $d_{k+1} > 0$ is assumed to be smaller than $(q_k - q_{k+1})/2^{k+1}$. We will impose another condition on d_{k+1}. Let $\dot{D}_{k+1,a,y}^{**} = \{(t,x) : t \ge 0, x < h_{k+1,a,y}^{**}(t)\}$, where $h_{k+1,a,y}^{**}(t)$ is the minimum of two functions: $h_k(t)$ and the linear function passing through the points $(1 - a, h_k(1 - a))$ and $(1, y)$. The corresponding reflected Brownian motion will be denoted $X_t^{**,k+1,a,y}$. We let a range between 0 and 1 and y range between 0 and $h_k(1 - a)$. By Corollary 10.14 and the continuity of probability, there exists a small $a_{k+1} > 0$ such that for all $a \in (0, a_{k+1})$, all $y \in (0, h_k(1 - a))$ and for all $j \le k$,

$$\mathbb{P}(X_1^{**,k+1,a,y} \in (-2d_j, 0)) \ge (1 - 2^{-k-1})\,\mathbb{P}(X_1^k \in (-2d_j, 0)). \qquad (10.9)$$

We choose $d_{k+1} > 0$ so small that $c_{k+1} < a_{k+1}$.

Recall the family of domains $\dot{D}_{1,y}^*$. We similarly define domains $\dot{D}_{k+1,y}^* = \{(t,x) : t \ge 0, x < h_{k+1,y}(t)\}$, where $h_{k+1,y}(t)$ is the minimum of two functions: $h_k(t)$ and the linear function passing through the points $(1 - c_{k+1}, h_k(1 - c_{k+1}))$ and $(1, y)$. Here y takes values in $[0, h_k(1 - c_{k+1})]$. As in the case of $\dot{D}_{1,0}^*$, the point $(1, 0)$ is not a heat atom in $\dot{D}_{k+1,0}^*$. By the same argument as in the case of $\dot{D}_{1,y}^*$, we can find $y_{k+1} \in (0, h_k(1 - c_{k+1}))$ such that $\mathbb{P}(X_1^{*,k+1,y_1} = 0) = q_{k+1}$. Then we let $h_{k+1}(t) = h_{k+1,y_{k+1}}(t)$.

Let $h_\infty(t) = \lim_{k\to\infty} h_k(t)$ and let $\dot{D}_\infty^*$ be the corresponding domain. Since h_k's converge uniformly to h_∞, Corollary 10.14 and the continuity of probability imply that $\mathbb{P}(X_1^\infty = 0) = \lim_{k\to\infty} q_k = q_\infty$. Depending on whether q_∞ is equal to 0 or not, we have an example of a domain $\dot{D}_\infty^*$ without or with a heat atom at $(1, 0)$. In order to prove parts (i) and (ii) of the theorem it remains to show that the heat equation solution $u_\infty(t, x)$ in $\dot{D}_\infty^*$ is singular at $(1, 0)$ for any value of q_∞.

Let A_k be the set in space-time consisting of two line segments, the first one, say I_k, joining the points $(1 - c_k, -d_k)$ and $(1 - c_k, h_k(1 - c_k))$, the other one, J_k, joining $(1 - c_k, -d_k)$ and $(1, -d_k)$.

Let T_k be the hitting time of A_k by the space-time process (t, X_t^k). Recall that we are assuming that all reflected Brownian motions are constructed from the same Brownian motion using the Skorohod Lemma. Since the space-time boundaries of

$\dot{D}^*_{k-1}$ and $\dot{D}^*_k$ agree except on a set which is blocked away from the line $\{(t, x) : t = 0\}$ by A_k, the hitting time of A_k by the space-time process (t, X^{k-1}_t) is the same as for (t, X^k_t), a.s. Recall that the size of the heat atom at $(1, 0)$ for the process X^{k-1}_t is equal to q_{k-1}. Hence, the process X^{k-1}_t has to hit A_k with probability greater than q_{k-1} and the earlier remark implies that the same is true for X^k_t. We will now estimate the probability that the process X^k_t hits A_k and then takes a value less than $-2d_k$ at time 1. Let S_k be the smallest $s \geq T_k$ such that $X^k_s \in J_k$. Note that if $T_k < \infty$ but $S_k = \infty$ then $X^k_1 \geq -2d_k$.

Let $\hat{X}_t$ be reflected Brownian motion in $\{(t, x) : x < 0\}$, starting from a point $(s, -d_k) \in J_k$. The probability that $\hat{X}_1 < -2d_k$ is less than the probability that the standard Brownian motion differs from its starting point at time c_k by more than d_k, which we have named p_k. By Corollary 10.13, for any $(s, -d_k) \in J_k$,

$$\mathbb{P}(X^k_1 < -2d_k \mid X^k_s = -d_k) \leq p_k.$$

Hence, the probability that X^k_t hits A_k and subsequently ends up below $-2d_k$ at time 1 is less than p_k. This implies that the probability that X^k_1 is strictly between $-2d_k$ and 0 is greater than $q_{k-1} - p_k - q_k$. By our choice of c_k, this probability is greater than $(q_{k-1} - q_k)/2$, i.e.,

$$\mathbb{P}(X^k_1 \in (-2d_k, 0)) \geq (q_{k-1} - q_k)/2. \tag{10.10}$$

In view of our assumption that $c_{k+1} < a_{k+1}$ and (10.9), we have for $k \geq j$,

$$\mathbb{P}(X^{k+1}_1 \in (-2d_j, 0)) \geq (1 - 2^{-k-1}) \mathbb{P}(X^k_1 \in (-2d_j, 0)).$$

Since $\prod_{k \geq 1}(1 - 2^{-k-1}) > 0$, combining (10.10) with these inequalities for all $k \geq j$ shows that we have for some absolute constant $\alpha > 0$, every j and $k \geq j$,

$$\mathbb{P}(X^{k+1}_1 \in (-2d_j, 0)) \geq \alpha(q_{j-1} - q_j)/2.$$

By passing to the limit, for every j,

$$\mathbb{P}(X^\infty_1 \in (-2d_j, 0)) \geq \alpha(q_{j-1} - q_j)/2.$$

By Theorem 10.17 below, $x \to u_\infty(1, x)$ is a non-decreasing function. Since $d_k \leq (q_{k-1} - q_k)/2^k$ and

$$\int_{-2d_k}^0 u_\infty(1, x)dx \geq \alpha(q_{k-1} - q_k)/2,$$

we must have $u_\infty(1, x) \geq 2^{k-2}\alpha$ for some $x \in (-2d_k, 0)$. By monotonicity,

$$\lim_{x \uparrow 0} u_\infty(1, x) = \infty. \qquad \square$$

We will use the standard notation $\Phi(a) = \int_{-\infty}^{a} \sqrt{1/2\pi} \exp(-y^2/2)dy$.

Lemma 10.16. *Suppose that X_t is a reflected Brownian motion in $\dot{D} = \{(t, y) : y \leq \theta t\}$, where $\theta > 0$ is a constant. If $X_0 = 0$ and L_s denotes the local time of X_t on the boundary of $\dot{D}$ then*

$$\mathbb{E}\exp(2\theta L_s) = \int_0^\infty \left\{ 2\theta\Phi\left(\frac{s\theta - m}{\sqrt{s}}\right) + \frac{2}{\sqrt{2\pi s}}\exp\left(-\frac{(s\theta - m)^2}{2s}\right)\right\} dm.$$

$$(10.11)$$

From this we obtain the following three estimates,

$$\mathbb{E}\exp(2\theta L_s) \leq 1 + 2\theta\sqrt{s}\int_{-\infty}^{\theta\sqrt{s}} \Phi(y)dy + \sqrt{\frac{2}{\pi}}\theta\sqrt{s}, \qquad (10.12)$$

$$\mathbb{E}\exp(2\theta L_s) \geq 1 + 2\theta\sqrt{s}\int_{-\infty}^{\theta\sqrt{s}} \Phi(y)dy, \qquad (10.13)$$

$$\mathbb{E}\exp(2\theta L_s) \leq 2 + 4s\theta^2 + \sqrt{2/\pi}\frac{1}{\theta\sqrt{s}}\exp(-\theta^2 s/2). \qquad (10.14)$$

The estimate (10.12) is useful for small values of $\theta\sqrt{s}$ while the one in (10.14) is useful when $\theta\sqrt{s}$ is large.

Proof. We start by finding a formula for the density of L_s, for a fixed s.

Let B_t^* denote Brownian motion with drift $-\theta$ on the half line $(-\infty, 0]$, reflected at 0, starting from $B_0^* = 0$. The random variable L_s has the same distribution as the amount of local time L_s^* spent by B_t^* at 0 before time s. Let $\tilde{B}$ be a Brownian motion with constant drift $-\theta$, no reflection, starting from $\tilde{B}_0 = 0$. By the Skorohod construction of reflected processes, the distribution of L_s^* is the same as the distribution of the maximum of $\tilde{B}_t$ over the interval $[0, s]$, so L_s is distributed as $\tilde{M}_s = \sup_{t\in[0,s]} \tilde{B}_t$.

Let $\hat{B}_t$ be the standard (non-reflected) Brownian motion with $\hat{B}_0 = 0$ and consider its maximum process $\hat{M}_s = \sup_{t\in[0,s]} \hat{B}_t$. By [KS91, Proposition 2.8.1],

$$\mathbb{P}(\hat{M}_s \in dm, \hat{B}_s \in db) = \frac{2(2m - b)}{\sqrt{2\pi s^3}}\exp\left(-\frac{(2m - b)^2}{2s}\right) dm\, db.$$

By Girsanov's theorem,

$$\mathbb{P}(\tilde{M}_s \in dm, \tilde{B}_s \in db) = \frac{2(2m - b)}{\sqrt{2\pi s^3}}\exp\left(-\frac{(2m - b)^2}{2s}\right)\exp\left(-\theta b - \frac{\theta^2 s}{2}\right) dm\, db.$$

Hence,

$$
\begin{aligned}
\mathbb{P}(\tilde{M}_s \in dm) &= \int_{-\infty}^{m} \frac{2(2m-b)}{\sqrt{2\pi s^3}} \exp\left(-\frac{(2m-b)^2}{2s}\right) \exp\left(-\theta b - \frac{\theta^2 s}{2}\right) db \\
&= \frac{2}{\sqrt{2\pi s^3}} \exp\left(-\frac{\theta^2 s}{2}\right) \int_{-\infty}^{m} (2m-b) \exp\left(-\frac{(2m-b)^2}{2s} - \theta b\right) db \\
&= \frac{2}{\sqrt{2\pi s^3}} \exp\left(-\frac{\theta^2 s}{2}\right) \int_{-\infty}^{m} (2m-b) \exp\left(-\frac{(b+s\theta-2m)^2}{2s}\right) \\
&\qquad\qquad \exp\left(-2\theta m + \frac{\theta^2 s}{2}\right) db \\
&= \frac{2}{\sqrt{2\pi s^3}} \exp(-2\theta m) \int_{-\infty}^{m} (2m-b) \exp\left(-\frac{(b+s\theta-2m)^2}{2s}\right) db \\
&= \frac{2}{\sqrt{2\pi s^3}} \exp(-2\theta m)s \int_{-\infty}^{m} \left(\theta - \frac{b+s\theta-2m}{s}\right) \\
&\qquad\qquad \exp\left(-\frac{(b+s\theta-2m)^2}{2s}\right) db \\
&= \frac{2}{\sqrt{2\pi s}} \exp(-2\theta m)\theta \int_{-\infty}^{m} \exp\left(-\frac{(b+s\theta-2m)^2}{2s}\right) db \\
&\quad + \frac{2}{\sqrt{2\pi s}} \exp(-2\theta m) \int_{-\infty}^{m} \left(-\frac{b+s\theta-2m}{s}\right) \\
&\qquad\qquad \exp\left(-\frac{(b+s\theta-2m)^2}{2s}\right) db \\
&= \frac{2}{\sqrt{2\pi s}} \exp(-2\theta m)\theta \int_{-\infty}^{(s\theta-m)/\sqrt{s}} \exp\left(-\frac{y^2}{2}\right) \sqrt{s}\, dy \\
&\quad + \frac{2}{\sqrt{2\pi s}} \exp(-2\theta m) \exp\left(-\frac{(b+s\theta-2m)^2}{2s}\right) \Big|_{b=-\infty}^{b=m} \\
&= \frac{2}{\sqrt{2\pi s}} \exp(-2\theta m)\theta \sqrt{2\pi s}\, \Phi\left(\frac{s\theta-m}{\sqrt{s}}\right) + \frac{2}{\sqrt{2\pi s}} \exp(-2\theta m) \\
&\qquad\qquad \exp\left(-\frac{(s\theta-m)^2}{2s}\right) \\
&= \exp(-2\theta m) \left\{ 2\theta\, \Phi\left(\frac{s\theta-m}{\sqrt{s}}\right) + \frac{2}{\sqrt{2\pi s}} \exp\left(-\frac{(s\theta-m)^2}{2s}\right) \right\}.
\end{aligned}
$$

This implies that

$\mathbb{E}\exp(2\theta L_s)$

$$= \int_0^\infty \exp(2\theta m)\exp(-2\theta m)\left\{2\theta\Phi\left(\frac{s\theta-m}{\sqrt{s}}\right)+\frac{2}{\sqrt{2\pi s}}\exp\left(-\frac{(s\theta-m)^2}{2s}\right)\right\}dm$$

$$= \int_0^\infty \left\{2\theta\Phi\left(\frac{s\theta-m}{\sqrt{s}}\right)+\frac{2}{\sqrt{2\pi s}}\exp\left(-\frac{(s\theta-m)^2}{2s}\right)\right\}dm,$$

and completes the proof of (10.11).

We will now prove (10.12). Change of variable $y=(s\theta-m)/\sqrt{s}$ yields

$$\int_0^\infty 2\theta\Phi\left(\frac{s\theta-m}{\sqrt{s}}\right)dm = \int_{\theta\sqrt{s}}^{-\infty}2\theta\Phi(y)(-dy\sqrt{s}) = 2\theta\sqrt{s}\int_{-\infty}^{\theta\sqrt{s}}\Phi(y)dy. \tag{10.15}$$

We also have

$$\int_0^\infty \frac{2}{\sqrt{2\pi s}}\exp\left(-\frac{(s\theta-m)^2}{2s}\right)dm = \left(\int_0^{s\theta}+\int_{s\theta}^\infty\right)\frac{2}{\sqrt{2\pi s}}\exp\left(-\frac{(s\theta-m)^2}{2s}\right)dm$$

$$\le s\theta\frac{2}{\sqrt{2\pi s}}+1 = 1+\sqrt{\frac{2}{\pi}}\theta\sqrt{s}. \tag{10.16}$$

The last two estimates combined with (10.11) give (10.12).

A calculation similar to that in (10.16) shows that

$$\int_0^\infty \frac{2}{\sqrt{2\pi s}}\exp\left(-\frac{(s\theta-m)^2}{2s}\right)dm \ge 1,$$

so this and (10.15) combined with (10.11) yield (10.13).

We turn to the proof of the estimate (10.14). Since

$$\int_{-\infty}^\infty \frac{1}{\sqrt{2\pi s}}\exp\left(-\frac{(s\theta-m)^2}{2s}\right)dm = 1,$$

we obtain from (10.11),

$$\mathbb{E}\exp(2\theta L_s) \le 2+\int_0^\infty 2\theta\Phi\left(\frac{s\theta-m}{\sqrt{s}}\right)dm$$

$$= 2+\int_0^{2s\theta}2\theta\Phi\left(\frac{s\theta-m}{\sqrt{s}}\right)dm+\int_{2s\theta}^\infty 2\theta\Phi\left(\frac{s\theta-m}{\sqrt{s}}\right)dm$$

$$\le 2+\int_0^{2s\theta}2\theta\,dm+\int_{2s\theta}^\infty 2\theta\Phi\left(\frac{s\theta-m}{\sqrt{s}}\right)dm$$

$$= 2+4s\theta^2+\int_{2s\theta}^\infty 2\theta\Phi\left(\frac{s\theta-m}{\sqrt{s}}\right)dm. \tag{10.17}$$

In order to estimate the integral in the last line we recall a standard inequality. For $a > 0$,

$$\int_a^\infty \exp(-y^2/2)dy \le \int_a^\infty (y/a)\exp(-y^2/2)dy = (1/a)\exp(-a^2/2),$$

so for $m \ge 2s\theta$,

$$\Phi\left(\frac{s\theta - m}{\sqrt{s}}\right) = \int_{(m-s\theta)/\sqrt{s}}^\infty (1/\sqrt{2\pi})\exp(-y^2/2)dy$$

$$\le \frac{\sqrt{s}}{(m - s\theta)\sqrt{2\pi}}\exp\left(-\frac{(s\theta - m)^2}{2s}\right).$$

This implies that

$$\int_{2s\theta}^\infty 2\theta\,\Phi\left(\frac{s\theta - m}{\sqrt{s}}\right)dm \le 2\theta\int_{2s\theta}^\infty \frac{\sqrt{s}}{(m - s\theta)\sqrt{2\pi}}\exp\left(-\frac{(m - s\theta)^2}{2s}\right)dm$$

$$= 2\theta\int_{\theta\sqrt{s}}^\infty \frac{\sqrt{s}}{y\sqrt{2\pi}}\exp(-y^2/2)dy.$$

For $y \ge \theta\sqrt{s}$ we have $1/y \le y/(\theta^2 s)$, so

$$\int_{2s\theta}^\infty 2\theta\,\Phi\left(\frac{s\theta - m}{\sqrt{s}}\right)dm \le 2\theta\int_{\theta\sqrt{s}}^\infty \frac{\sqrt{s}\,y}{\theta^2 s\sqrt{2\pi}}\exp(-y^2/2)dy$$

$$= \sqrt{2/\pi}\,\frac{1}{\theta\sqrt{s}}\exp(-\theta^2 s/2).$$

We combine this estimate with (10.17) to conclude that

$$\mathbb{E}\exp(2\theta L_s) \le 2 + 4s\theta^2 + \sqrt{2/\pi}\,\frac{1}{\theta\sqrt{s}}\exp(-\theta^2 s/2). \qquad \square$$

Proof of Theorem 10.10. The claims about existence or non-existence of heat atoms follow easily from the Kolmogorov criterion (10.1). It remains to analyze existence of singularities of $u(t, x)$.

(i) The absence of a singularity at $t = 1$ under the assumptions of part (i) is a direct consequence of Proposition 10.11, proved below.

(ii) We will show the required property for domain $\dot{D} = \{(t, x) : t > 0, x < g(t)\}$ where g satisfies the following conditions: $g(t) = 0$ for $t \ge 1$; g is smooth on $[0, 1)$; $g(1 - t) = \sqrt{t}|\log t|^\beta$ for some $-1 \le \beta < 0$ and all $t \in [0, 2^{-k_0})$; g is constant on $[0, 1 - 2^{-k_0+1}]$; g is decreasing and concave on $(0, 1)$. Here k_0 is a large positive integer. In view of Lemma 10.8 we can and will assume that $u(0, x) \equiv 1$; we need this assumption to be able to apply Theorem 10.15. Recall the notation from Theorem 10.15. Let $v(t, x) = u(t, g(t) + x)$. By Theorem 10.15, the integration by parts formula and Jensen's inequality, we have for $t < 1$ and $x < 0$,

$$
\begin{aligned}
v(t, x) &= \exp\left(-\frac{1}{2}\int_0^t |g'(s)|^2 ds\right) \mathbb{E}_x\left[\exp\left(\int_0^t B_s g''(t - s)ds - 2\int_0^t L_s g''(t - s)ds\right)\right] \\
&\geq \exp\left(-\frac{1}{2}\int_0^t |g'(s)|^2 ds\right) \exp\left(-2\int_0^t \mathbb{E}_x[L_s] g''(t - s)ds\right) \\
&= \exp\left(\int_0^t (-\frac{1}{2}|g'(s)|^2 - 2g''(s)\,\mathbb{E}_x[L_{t-s}])ds\right).
\end{aligned}
\tag{10.18}
$$

Since

$$
\mathbb{E}_x[L_s] = \int_0^s \frac{2}{\sqrt{2\pi\theta}} e^{-x^2/(2\theta)} d\theta = \sqrt{2/\pi}\, s^{1/2} \int_0^1 \frac{1}{\sqrt{r}} e^{-x^2/(2sr)} dr,
$$

we have

$$
\mathbb{E}_{\lambda\sqrt{\varepsilon}}[L_s] \geq \sqrt{2/\pi}\, s^{1/2} \int_0^1 e^{-\lambda^2/r} r^{-1/2} dr \quad \text{for } \lambda > 0.
\tag{10.19}
$$

It is straightforward to check that as $t \uparrow 1$,

$$
g'(t) = -\frac{1}{2}(1 - t)^{-1/2} |\log(1 - t)|^\beta (1 + o(1 - t)),
$$

$$
g''(t) = -\frac{1}{4}(1 - t)^{-3/2} |\log(1 - t)|^\beta (1 + o(1 - t)).
$$

Totally elementary but somewhat tedious calculations combining these estimates with (10.18) and (10.19) yield for $-1 \leq \beta < 0$ and $\lambda > 0$ small enough,

$$
\lim_{\varepsilon \to 0} \log v(1 - \varepsilon, -\lambda\sqrt{\varepsilon}) = \infty.
$$

This implies that

$$
\lim_{\varepsilon \to 0} u(1 - \varepsilon, g(1 - \varepsilon) - \lambda\sqrt{\varepsilon}) = \infty.
$$

If $-1 \leq \beta < 0$ then $g(1 - \varepsilon) - \lambda\sqrt{\varepsilon} < g(1)$ for small $\varepsilon > 0$. We have assumed that g is decreasing and concave, so by Theorem 10.18 below,

$$
u(1, g(1 - \varepsilon) - \lambda\sqrt{\varepsilon}) \geq u(1 - \varepsilon, g(1 - \varepsilon) - \lambda\sqrt{\varepsilon}).
$$

Thus $\lim_{\varepsilon \to 0} u(1, g(1-\varepsilon) - \lambda\sqrt{\varepsilon}) = \infty$. Since $g(1-\varepsilon) - \lambda\sqrt{\varepsilon} \to g(1)$ as $\varepsilon \to 0$, we see that for $-1 \le \beta < 0$, $u(1, x)$ is singular at $g(1) = 0$.

(iii) The following is a special case of a probabilistic representation of $u(1, x)$ given in [BCS03, Corollary 2.12]. Let $\tilde{D} = \{(t, x) : (1-t, x) \in \dot{D}\}$, $f(t) = g(1-t)$, and let Y_t be the reflected Brownian motion in $\tilde{D}$ with $Y_0 = x$, for some fixed $x < g(1)$. We will denote the local time spent by Y_t on the boundary of $\tilde{D}$ by L_s. Then

$$u(1, x) = \mathbb{E}\left[\exp\left(\int_0^1 2f'(t)dL_t\right) u(0, Y_1)\right].$$

We can assume without loss of generality that $u(0, x) \equiv 1$. We have $g'(t) = 0$ for $t \in (0, 1 - 2^{-k_0})$ so

$$u(1, x) \le \mathbb{E}\exp\left(\int_0^{2^{-k_0}} 2f'(t)dL_t\right).$$

Let M_k be the line segment in time-space which connects $(2^{-k}, f(2^{-k}))$ and $(2^{-k+1}, f(2^{-k+1}))$. Its slope θ_k is equal to

$$\frac{f(2^{-k+1}) - f(2^{-k})}{2^{-k}} = \frac{\sqrt{2^{-k+1}}(k-1)^\beta - \sqrt{2^{-k}}k^\beta}{2^{-k}} = 2^{k/2}(\sqrt{2}(k-1)^\beta - k^\beta).$$

For arbitrary $c_1 > 1$, this is bounded above by $c_1 2^{k/2}(\sqrt{2} - 1)k^\beta$ and it is bounded below by $(1/c_1)2^{k/2}(\sqrt{2} - 1)k^\beta$, for sufficiently large k. Let N_k be the straight line which contains M_k. The line N_k intersects the t-axis at the point $f(2^{-k}) - \theta_k 2^{-k}$ which for large k is greater than $\sqrt{2^{-k}}k^\beta - c_1 2^{k/2}(\sqrt{2} - 1)k^\beta 2^{-k}$. We can take any $c_1 > 1$, so we let $1 - c_1(\sqrt{2} - 1) = 1/2$. Then we see that N_k crosses the t-axis at a distance d_k from 0 which is greater than $(1/2)\sqrt{2^{-k}}k^\beta$, for large k. Let B_t be a Brownian motion starting from $B_0 = 0$. The probability that B_t ever hits N_k is equal to $\exp(-2d_k\theta_k)$, see, e.g., [KS91, (5.13)]. Hence, the probability of hitting N_k by B_t is bounded by

$$\exp(-2(1/2)\sqrt{2^{-k}}k^\beta(1/c_1)2^{k/2}(\sqrt{2} - 1)k^\beta) = \exp(-(1/c_1)(\sqrt{2} - 1)k^{2\beta}).$$

We can suppose that Y_t is constructed from B_t using Lemma 10.12. Hence, the probability that Y_t ever hits N_k is bounded by the same quantity. The probability of hitting M_k by Y_t is even smaller. The chance that Y_t hits $f(t)$ at some time $t \in [2^{-k}, 2^{-k+1}]$ is smaller again.
Let $T_k = \inf\{t \in [2^{-k}, 2^{-k+1}] : Y_t = f(t)\}$ and $A_k = \{T_k < \infty\}$. Then

$$\mathbb{P}(A_k) \le \exp(-(1/c_1)(\sqrt{2} - 1)k^{2\beta}).$$

We have

$$u(1,x) \leq \mathbb{E}\exp\left(\int_0^{2^{-k_0}} 2f'(t)dL_t\right) \leq \sum_{k=k_0+1}^{\infty} \mathbb{E}\left[\mathbf{1}_{A_k}\exp\left(\int_{T_k}^{2^{-k_0}} 2f'(t)dL_t\right)\right].$$

Suppose that $T_k < \infty$ so that $T_k \in [2^{-k}, 2^{-k+1})$. The derivative θ of $f(t)$ is bounded above by $c_2 2^{k/2}k^\beta$ for $t \in [2^{-k}, 2^{-k+1}]$, where $c_2 = c_2(\beta)$ does not depend on k. The length of the interval $[T_k, 2^{-k+1}]$ is not greater than 2^{-k}.

Note that for large $s\theta^2$, the estimate (10.14) yields $\mathbb{E}\exp(2\theta L_s) \leq 5s\theta^2$. We will apply (10.14) with θ replaced by $c_2 2^{k/2}k^\beta$ and $s = 2^{-k}$. Note that $\theta^2 s \leq (c_2 2^{k/2}k^\beta)^2 2^{-k} = c_2^2 k^{2\beta}$. Hence,

$$\mathbb{E}\exp\left(\int_{T_k}^{2^{-k+1}} 2f'(Y_s)dL_s\right) \leq c_3 k^{2\beta}.$$

In fact, we have proved a somewhat stronger assertion,

$$\mathbb{E}\left(\exp\left(\int_{2^{-k}}^{2^{-k+1}} 2f'(Y_s)dL_s\right)\Big|\mathcal{F}_{2^{-k}}\right) = \mathbb{E}\left(\exp\left(\int_{T_k}^{2^{-k+1}} 2f'(Y_s)dL_s\right)\Big|\mathcal{F}_{2^{-k}}\right)$$

$$\leq c_3 k^{2\beta},$$

where $\mathcal{F}_{2^{-k}} = \sigma\{Y_t : t \leq 2^{-k}\}$. By the strong Markov property,

$$\mathbb{E}\left[\mathbf{1}_{A_k}\exp\left(\int_{T_k}^{2^{-k_0}} 2f'(t)dL_t\right)\right] \leq \mathbb{E}\left[\mathbf{1}_{A_k}\exp\left(\sum_{j=k_0+1}^{k}\int_{2^{-j}}^{2^{-j+1}} 2f'(t)dL_t\right)\right]$$

$$\leq \mathbb{P}(A_k)\prod_{j=k_0+1}^{k} c_3 j^{2\beta}.$$

From Stirling's formula we obtain for large k,

$$\prod_{j=k_0+1}^{k} c_3 j^{2\beta} \leq c_3^{k-1}(k!)^{2\beta} \leq c_4 c_3^{k-1}(k^k e^{-k}\sqrt{2\pi k})^{2\beta}$$

$$= \exp(\log c_4 + (k-1)\log c_3 + 2\beta k \log k$$

$$-2\beta k + \beta \log(2\pi) + \beta \log k)$$

$$\leq \exp(4\beta k \log k),$$

so, using the assumption that $\beta > 1/2$ we obtain for large k,

$$\mathbb{E}\left[\mathbf{1}_{A_k} \exp\left(\int_{T_k}^{1/2} 2f'(t)dL_t\right)\right] \le \mathbb{P}(A_k)\exp(4\beta k \log k)$$

$$\le \exp(-(1/c_1)(\sqrt{2}-1)k^{2\beta})\exp(4\beta k \log k)$$

$$\le \exp(-c_5 k^{2\beta}).$$

Hence,

$$u(1,x) \le \sum_{k=k_0+1}^{\infty} \mathbb{E}\left[\mathbf{1}_{A_k} \exp\left(\int_{T_k}^{2^{-k_0}} 2f'(t)dL_t\right)\right] \le \sum_{k=k_0+1}^{\infty} \exp(-c_5 k^{2\beta}) < \infty.$$

Since the constant c_5 does not depend on x, the point $(1, g(1))$ is not a singularity for $u(t,x)$. □

Proof of Proposition 10.11. Non-existence of a heat atom follows from [BCS04, Theorem 3.6] and the lack of heat atoms for reflected Brownian motion on the boundary of a half-line.

We will apply the representation of the heat equation solution given in Theorem 10.15. Suppose $x < 0$ and let $Y = x + B_t - L_t$ be the reflected Brownian motion on $(-\infty, 0]$. Letting $v(t,x) = u(t, g(t) + x)$, we have by (10.4),

$$v(1,x) = \exp\left(-\frac{1}{2}\int_0^1 |g'(s)|^2 ds\right)\mathbb{E}\left[\exp\left(\int_0^1 g'(1-s)dB_s - 2\int_0^1 g'(1-s)dL_s\right)\right].$$

By the Cauchy-Schwartz inequality,

$$v(1,x) \le \exp\left(\frac{1}{2}\int_0^1 |g'(s)|^2 ds\right) \tag{10.20}$$

$$\times \sqrt{\mathbb{E}_x\left[\exp\left(2\int_0^1 g'(1-s)dB_s\right)\right]}\sqrt{\mathbb{E}_x\left[\exp 4\int_0^1 |g'(1-s)|dL_s\right]}.$$

All we have to do is to find a bound for each factor on the right hand side of (10.20), independent of x. The first factor, $\exp[(1/2)\int_0^1 |g'(s)|^2 ds]$, is bounded by assumption. We use [BCS04, (3.10)] to bound the second factor,

$$\mathbb{E}\exp\left(\int_0^1 2(g_n' - g')(1-s)dB_s\right) = \exp\left(\int_0^1 2|g_n' - g'|^2(s)ds\right) < \infty.$$

Let $p_t(x,y)$ be the transition density function for reflected Brownian motion Y on $(-\infty, 0]$. It is well known that

$$p_t(x, y) = \frac{1}{\sqrt{2\pi t}} \left(e^{-(x-y)^2/2t} + e^{-(x+y)^2/2t} \right).$$

Since singularity is a local property, we may assume without loss of generality that

$$\int_0^1 |g'(1-s)| s^{-1/2} ds < \frac{\sqrt{2\pi}}{64}.$$

Then

$$\mathbb{E}_x \left[4 \int_0^1 |g'(1-s)| dL_s \right] = 4 \int_0^1 |g'(1-s)| p_s(x, 0) ds$$

$$\leq \frac{8}{\sqrt{2\pi}} \int_0^1 |g'(1-s)| s^{-1/2} ds < 1/8,$$

and hence by Khasminskii's inequality

$$\mathbb{E}_x \left[\exp \left(4 \int_0^1 |g'(1-s)| dL_s \right) \right] \leq \frac{1}{1 - \frac{1}{2}} = 2.$$

This gives a bound for the last factor in (10.20) and completes the proof.   □

10.3 Monotonicity Properties of the Heat Equation Solution

Theorem 10.17. *Suppose that* $\dot{D} = \{(t, x) : t \geq 0, g_1(t) < x < g_2(t)\}$, *where* g_1 *and* g_2 *are continuous non-decreasing functions satisfying* $g_1(t) < g_2(t)$ *for every* t. *If* $x \to u(0, x)$ *is non-increasing then for every fixed* $t > 0$, *the function* $x \to u(t, x)$ *is non-increasing.*

See [BCS03] for a rather easy probabilistic proof of Theorem 10.17. Theorem 10.17 can be also easily proved using the maximum principle for parabolic functions (that is, without using probability) but we do not know a purely analytic proof of Theorem 10.18 below.

Theorem 10.18. *Suppose that* $\dot{D} = \{(t, x) : t \geq 0, x < g(t)\}$, *where* $g(t)$ *is a continuous decreasing concave function on* $[0, t_*]$. *If* $u(0, x) \equiv 1$ *then for any fixed* $x \leq g(t_*)$, *the function* $t \to u(t, x)$ *is non-decreasing on* $[0, t_*]$.

Remark 10.19.   (i) Theorem 10.18 holds also for domains of the form $\dot{D} = \{(t, x) : t \geq 0, g_1(t) < x < g_2(t)\}$, where $g_1(t)$ is non-decreasing and convex but $g_2(t)$ is non-increasing and concave. The proof is an elementary generalization of the proof of Theorem 3.2 and so it is left to the reader.

(ii) The assumption of concavity cannot be dropped. To see that monotonicity of $g(t)$ is not enough, consider the domain corresponding to the function $g(t)$ which is piecewise linear, with vertices at $(0, 1), (1 - \varepsilon, 1), (1, \varepsilon)$ and (∞, ε). If $\varepsilon > 0$ is very small then it is easy to see that $u(t, 0)$ is very large for $t > 1$ but close to 1. Since $u(0, 0) = 1$ and $u(t, 0) \to 1$ as $t \to \infty$, we see that $t \to u(t, 0)$ is not monotone.

(iii) The statement that $t \to u(t, x)$ is increasing for all t and x is equivalent to the assertion that $x \to u(t, x)$ is concave for all t and x. It is natural to ask whether the assumption in Theorem 10.18 that $u(0, x) \equiv 1$ could be relaxed and replaced with the assumption that $x \to u(0, x)$ is concave for all x. The answer is no—we leave an easy counterexample to the reader.

(iv) The function $t \to u(t, x)$ need not be decreasing when $g(t)$ is increasing and convex, for some functions $g(t)$ and some $x > g(0)$. We could not resolve the problem for $x < g(0)$—perhaps for such x, the function $t \to u(t, x)$ has to be decreasing when $g(t)$ is increasing and convex.

Proof of Theorem 10.18. Assume that $g(t)$ is smooth. Fix an $x < g(t_*)$ and any $0 \leq t_1 \leq t_2 \leq t_*$. Let $f_k(t) = g(t_k - t)$ for $t \in [0, t_1]$ and $k = 1, 2$. Let $\tilde{D}_k = \{(t, x) : x < f_k(t)\}$ and let X_t^k be the reflected Brownian motion in $\tilde{D}_k$, starting from $X_0^k = x$. Let L_t^k denote the local time spent by X_t^k on the boundary of $\tilde{D}_k$. We will assume that X_t^1 and X_t^2 are constructed from the same Brownian motion B_t via Lemma 10.12. Since $f_2(t) \leq f_1(t)$ for all $t \in [0, t_1]$, we have $L_t^2 \geq L_t^1$ for $t \in [0, t_1]$, by Corollary 10.13. Note that, by assumption of concavity,

$$f_2'(s_1) \geq f_2'(s_2) \geq f_1'(s_2), \tag{10.21}$$

for $s_1, s_2 \in [0, t_1]$ such that $s_1 \leq s_2$. We will show that these observations imply

$$\int_0^{t_2} f_2'(t) dL_t^2 \geq \int_0^{t_1} f_2'(t) dL_t^2 \geq \int_0^{t_1} f_1'(t) dL_t^1. \tag{10.22}$$

The first inequality is trivial; the second one is a simple measure-theoretical fact which we will prove using probabilistic ideas. Recall that $L_0^1 = L_0^2 = 0$ and define two probability measures K^1 and K^2 on $[0, t_1]$ as follows,

$$K^2[s_1, s_2] = (L_{s_2}^2 - L_{s_1}^2)/L_{t_1}^2, \qquad \text{for } 0 \leq s_1 \leq s_2 \leq t_1,$$

$$K^1[s_1, s_2] = (L_{s_2}^1 - L_{s_1}^1)/L_{t_1}^2, \qquad \text{for } 0 \leq s_1 \leq s_2 \leq t_1,$$

$$K^1(\{t_1\}) = (L_{t_1}^2 - L_{t_1}^1)/L_{t_1}^2.$$

Let $h(t) = \inf\{s : K^1[0, s] \geq K^2[0, t]\}$ and note that $h(t) \geq t$ for all $t \in [0, t_1]$ because $L_t^2 \geq L_t^1$. If V is a random variable with distribution K^2 then $h(V)$ has the distribution K^1. In view of monotonicity of h and (10.21), we have $f_2'(V) \geq f_1'(h(V))$. Hence,

$$\int_0^{t_1} f_2'(t)dL_t^2 = L_{t_1}^2 \ \mathbb{E}[f_2'(V)] \geq L_{t_1}^2 \ \mathbb{E}[f_1'(h(V))] \geq \int_0^{t_1} f_1'(t)dL_t^1.$$

We now apply (10.22) to obtain

$$u(t_2, x) = \mathbb{E}_x \left[\exp \left(\int_0^{t_2} 2f_2'(s)dL_s^2 \right) u(0, X_{t_2}^2) \right]$$

$$= \mathbb{E}_x \exp \left(\int_0^{t_2} 2f_2'(s)dL_s^2 \right) \geq \mathbb{E}_x \exp \left(\int_0^{t_1} 2f_1'(s)dL_s^1 \right) = u(t_1, x).$$

The proof for non-smooth $g(t)$ can be obtained by approximating the domain with domains with smooth decreasing concave boundaries.   □

References

[AB02]   R. Atar, K. Burdzy, On nodal lines of Neumann eigenfunctions. Electron. Commun.
            Probab. **7**, 129–139 (2002)
[AB04]   R. Atar, K. Burdzy, On Neumann eigenfunctions in lip domains. J. Am. Math. Soc.
            17(2), 243–265 (2004)
[ACTDBP06] N. Arcozzi, E. Casadio Tarabusi, F. Di Biase, M.A. Picardello, Twist points of
            planar domains. Trans. Am. Math. Soc. **358**(6), 2781–2798 (2006)
[Ahl78]   L.V. Ahlfors, *Complex Analysis. An Introduction to the Theory of Analytic
            Functions of One Complex Variable*. International Series in Pure and Applied
            Mathematics, 3rd edn. (McGraw-Hill Book, New York, 1978)
[Ata01]   R. Atar, Invariant wedges for a two-point reflecting Brownian motion and the "hot
            spots" problem. Electron. J. Probab. **6**(18), 19 pp. (2001)
[Ath00]   S. Athreya, Monotonicity property for a class of semilinear partial differential
            equations, in *Séminaire de Probabilités, XXXIV*. Lecture Notes in Mathematics,
            vol. 1729 (Springer, Berlin, 2000), pp. 388–392
[Ban80]   C. Bandle, *Isoperimetric Inequalities and Applications*. Monographs and Studies
            in Mathematics, vol. 7 (Pitman (Advanced Publishing Program), Boston, 1980)
[Bañ87]   R. Bañuelos, On an estimate of Cranston and McConnell for elliptic diffusions in
            uniform domains. *Probab. Theory Relat. Fields* **76**(3), 311–323 (1987)
[Bas95]   R.F. Bass, *Probabilistic Techniques in Analysis*. Probability and Its Applications
            (Springer, New York, 1995)
[BB92]   R.F. Bass, K. Burdzy, Lifetimes of conditioned diffusions. Probab. Theory Relat.
            Fields **91**(3–4), 405–443 (1992)
[BB99]   R. Bañuelos, K. Burdzy, On the "hot spots" conjecture of J. Rauch. J. Funct. Anal.
            164(1), 1–33 (1999)
[BB00]   R.F. Bass, K. Burdzy, Fiber Brownian motion and the "hot spots" problem. Duke
            Math. J. **105**(1), 25–58 (2000)
[BB08]   R.F. Bass, K. Burdzy, On pathwise uniqueness for reflecting Brownian motion in
            $C^{1+\gamma}$ domains. Ann. Probab. **36**(6), 2311–2331 (2008)
[BBC05]   R.F. Bass, K. Burdzy, Z.-Q. Chen, Uniqueness for reflecting Brownian motion in
            lip domains. Ann. Inst. H. Poincaré Probab. Stat. **41**(2), 197–235 (2005)
[BC98]   K. Burdzy, Z.-Q. Chen, Weak convergence of reflecting Brownian motions.
            Electron. Commun. Probab. **3**, 29–33 (1998)
[BC02]   K. Burdzy, Z.-Q. Chen, Coalescence of synchronous couplings. Probab. Theory
            Relat. Fields **123**(4), 553–578 (2002)
[BCS03]   K. Burdzy, Z.-Q. Chen, J. Sylvester, The heat equation and reflected Brownian
            motion in time-dependent domains. II. Singularities of solutions. J. Funct. Anal.
            204(1), 1–34 (2003)

K. Burdzy, *Brownian Motion and its Applications to Mathematical Analysis*,   133
Lecture Notes in Mathematics 2106, DOI 10.1007/978-3-319-04394-4,
© Springer International Publishing Switzerland 2014

[BCS04]   K. Burdzy, Z.-Q. Chen, J. Sylvester, The heat equation and reflected Brownian
           motion in time-dependent domains. Ann. Probab. **32**(1B), 775–804 (2004)
[Ber96]   J. Bertoin, *Lévy Processes*. Cambridge Tracts in Mathematics, vol. 121 (Cambridge
           University Press, Cambridge, 1996)
[BH91]   R.F. Bass, P. Hsu, Some potential theory for reflecting Brownian motion in Hölder
           and Lipschitz domains. Ann. Probab. **19**(2), 486–508 (1991)
[Bil99]   P. Billingsley, *Convergence of Probability Measures*. Wiley Series in Probability
           and Statistics: Probability and Statistics, 2nd edn. (Wiley, New York, 1999)
[BK98]   K. Burdzy, D. Khoshnevisan, Brownian motion in a Brownian crack. Ann. Appl.
           Probab. **8**(3), 708–748 (1998)
[BK00]   K. Burdzy, W.S. Kendall, Efficient Markovian couplings: examples and counterex-
           amples. Ann. Appl. Probab. **10**(2), 362–409 (2000)
[BP04]   R. Bañuelos, M. Pang, An inequality for potentials and the "hot-spots" conjecture.
           Indiana Univ. Math. J. **53**(1), 35–47 (2004)
[BPP04]   R. Bañuelos, M. Pang, M. Pascu, Brownian motion with killing and reflection and
           the "hot-spots" problem. Probab. Theory Relat. Fields **130**(1), 56–68 (2004)
[Bry95]   W. Bryc, *The Normal Distribution*. Lecture Notes in Statistics, vol. 100 (Springer,
           New York, 1995). Characterizations with applications
[Bur76]   D.L. Burkholder, Harmonic analysis and probability, in *Studies in Harmonic
           Analysis (Proc. Conf., DePaul Univ., Chicago, Ill., 1974)*. MAA Stud. Math., vol.
           13 (Mathematical Association of America, Washington, 1976), pp. 136–149
[Bur90]   K. Burdzy, Minimal fine derivatives and Brownian excursions. Nagoya Math. J.
           119, 115–132 (1990)
[Bur05]   K. Burdzy, The hot spots problem in planar domains with one hole. Duke Math. J.
           129(3), 481–502 (2005)
[BW99]   K. Burdzy, W. Werner, A counterexample to the "hot spots" conjecture. Ann. Math.
           (2) **149**(1), 309–317 (1999)
[Cha84]   I. Chavel, *Eigenvalues in Riemannian Geometry*. Pure and Applied Mathematics,
           vol. 115 (Academic, Orlando, 1984). Including a chapter by Burton Randol, With
           an appendix by Jozef Dodziuk
[Che92a]   M.F. Chen, *From Markov Chains to Nonequilibrium Particle Systems* (World
           Scientific Publishing, River Edge, 1992)
[Che92b]   Z.Q. Chen, Pseudo Jordan domains and reflecting Brownian motions. Probab.
           Theory Relat. Fields **94**(2), 271–280 (1992)
[Chu84]   K.L. Chung, The lifetime of conditional Brownian motion in the plane. Ann. Inst.
           H. Poincaré Probab. Stat. **20**(4), 349–351 (1984)
[Dav75]   B. Davis, Picard's theorem and Brownian motion. Trans. Am. Math. Soc. **213**,
           353–362 (1975)
[Dav79a]   B. Davis, Applications of the conformal invariance of Brownian motion, in
           *Proceedings of the Symposium in Pure Mathematics of the American Mathematical
           Society*, Williams College, Williamstown, MA, 10–28 July, 1978, ed. by G. Weiss,
           S. Wainger. Harmonic Analysis in Euclidean Spaces. Part 2, vol. XXXV (American
           Mathematical Society, Providence, 1979), pp. 303–310. Dedicated to Nestor
           M. Rivière
[Dav79b]   B. Davis, Brownian motion and analytic functions. Ann. Probab. **7**(6), 913–932
           (1979)
[Dav89]   E.B. Davies, *Heat Kernels and Spectral Theory*. Cambridge Tracts in Mathematics,
           vol. 92 (Cambridge University Press, Cambridge, 1989)
[Doo61]   J.L. Doob, Conformally invariant cluster value theory. Ill. J. Math. **5**, 521–549
           (1961)
[Doo84]   J.L. Doob, *Classical Potential Theory and Its Probabilistic Counterpart*.
           Grundlehren der Mathematischen Wissenschaften (Fundamental Principles of
           Mathematical Sciences), vol. 262 (Springer, New York, 1984)

[Dur83]   P.L. Duren, *Univalent Functions.* Grundlehren der Mathematischen Wissenschaften (Fundamental Principles of Mathematical Sciences), vol. 259 (Springer, New York, 1983)

[Dur84]   R. Durrett, *Brownian Motion and Martingales in Analysis.* Wadsworth Mathematics Series (Wadsworth International Group, Belmont, 1984)

[Fit89]   P.J. Fitzsimmons, Time changes of symmetric Markov processes and a Feynman-Kac formula. J. Theor. Probab. **2**(4), 487–501 (1989)

[Fol76]   G.B. Folland, *Introduction to Partial Differential Equations* (Princeton University Press, Princeton, 1976). Preliminary informal notes of university courses and seminars in mathematics, Mathematical Notes

[FŌT94]   M. Fukushima, Y. Ōshima, M. Takeda, *Dirichlet Forms and Symmetric Markov Processes.* de Gruyter Studies in Mathematics, vol. 19 (Walter de Gruyter & Co., Berlin, 1994)

[Fre02]   P. Freitas, Closed nodal lines and interior hot spots of the second eigenfunction of the Laplacian on surfaces. Indiana Univ. Math. J. **51**(2), 305–316 (2002)

[Fuk67]   M. Fukushima, A construction of reflecting barrier Brownian motions for bounded domains. Osaka J. Math. **4**, 183–215 (1967)

[Gru98]   J.-C. Gruet, On the length of the homotopic Brownian word in the thrice punctured sphere. Probab. Theory Relat. Fields **111**(4), 489–516 (1998)

[HSS91]   R. Hempel, L.A. Seco, B. Simon, The essential spectrum of Neumann Laplacians on some bounded singular domains. J. Funct. Anal. **102**(2), 448–483 (1991)

[IM74]   K. Itô, H.P. McKean Jr., *Diffusion Processes and Their Sample Paths* (Springer, Berlin, 1974). Second printing, corrected, Die Grundlehren der mathematischen Wissenschaften, Band 125

[IM03]   K. Ishige, N. Mizoguchi, Location of blow-up set for a semilinear parabolic equation with large diffusion. Math. Ann. **327**(3), 487–511 (2003)

[IW89]   N. Ikeda, S. Watanabe, *Stochastic Differential Equations and Diffusion Processes.* North-Holland Mathematical Library, 2nd edn., vol. 24 (North-Holland, Amsterdam, 1989)

[Jer00]   D. Jerison, Locating the first nodal line in the Neumann problem. Trans. Am. Math. Soc. **352**(5), 2301–2317 (2000)

[JN00]   D. Jerison, N. Nadirashvili, The "hot spots" conjecture for domains with two axes of symmetry. J. Am. Math. Soc. **13**(4), 741–772 (2000)

[Kak44]   S. Kakutani, Two-dimensional Brownian motion and harmonic functions. Proc. Imp. Acad. Tokyo **20**, 706–714 (1944)

[Kak45a]   S. Kakutani, Markoff process and the Dirichlet problem. Proc. Jpn. Acad. **21**(1945), 227–233 (1949)

[Kak45b]   S. Kakutani, Two-dimensional Brownian motion and the type problem of Riemann surfaces. Proc. Jpn. Acad. **21**(1945), 138–140 (1949)

[Kaw85]   B. Kawohl, *Rearrangements and Convexity of Level Sets in PDE.* Lecture Notes in Mathematics, vol. 1150 (Springer, Berlin, 1985)

[Kni81]   F.B. Knight, *Essentials of Brownian Motion and Diffusion.* Mathematical Surveys, vol. 18 (American Mathematical Society, Providence, 1981)

[KS91]   I. Karatzas, S.E. Shreve, *Brownian Motion and Stochastic Calculus.* Graduate Texts in Mathematics, 2nd edn., vol. 113 (Springer, New York, 1991)

[Lin87]   C.S. Lin, On the second eigenfunctions of the Laplacian in $\mathbf{R}^2$. Commun. Math. Phys. **111**(2), 161–166 (1987)

[Lin92]   T. Lindvall, *Lectures on the Coupling Method.* Wiley Series in Probability and Mathematical Statistics: Probability and Mathematical Statistics (Wiley, New York, 1992)

[LS84]   P.-L. Lions, A.-S. Sznitman, Stochastic differential equations with reflecting boundary conditions. Commun. Pure Appl. Math. **37**(4), 511–537 (1984)

[Luo95]   S.L. Luo, A probabilistic proof of the fundamental theorem of algebra and a generalization. Math. Appl. (Wuhan) **8**(4), 487–489 (1995)

[McM69]   J.E. McMillan, Boundary behavior of a conformal mapping. Acta Math. **123**, 43–67 (1969)

[Miy09]   Y. Miyamoto, The "hot spots" conjecture for a certain class of planar convex domains. J. Math. Phys. **50**(10), 103530–103530-7 (2009)

[Mou88]   T.S. Mountford, Transience of a pair of local martingales. Proc. Am. Math. Soc. **103**(3), 933–938 (1988)

[MP10]   P. Mörters, Y. Peres, *Brownian Motion*. Cambridge Series in Statistical and Probabilistic Mathematics (Cambridge University Press, Cambridge, 2010). With an appendix by Oded Schramm and Wendelin Werner

[Nad86]   N.S. Nadirashvili, Multiplicity of eigenvalues of the Neumann problem. Dokl. Akad. Nauk SSSR **286**(6), 1303–1305 (1986)

[Nad87]   N.S. Nadirashvili, Multiple eigenvalues of the Laplace operator. Mat. Sb. (N.S.) **133**(175)(2), 223–237, 272 (1987)

[Nov81]   A.A. Novikov, Small deviations of Gaussian processes. Mat. Zametki **29**(2), 291–301, 319 (1981)

[NTY01]   N. Nadirashvili, D. Tot, D. Yakobson, Geometric properties of eigenfunctions. Uspekhi Mat. Nauk **56**(6(342)), 67–88 (2001)

[O'N11]   M.D. O'Neill, A Green proof of Fatou's theorem. J. Stat. Theory Pract. **5**(3), 497–513 (2011)

[O'N12]   M.D. O'Neill, A geometric and stochastic proof of the twist point theorem. Publ. Mat. **56**(1), 41–63 (2012)

[Pas02]   M.N. Pascu, Scaling coupling of reflecting Brownian motions and the hot spots problem. Trans. Am. Math. Soc. **354**(11), 4681–4702 (2002)

[Pas05]   M.N. Pascu, A probabilistic proof of the fundamental theorem of algebra. Proc. Am. Math. Soc. **133**(6), 1707–1711 (2005)

[PG11]   M.N. Pascu, M.E. Gageonea, Monotonicity properties of the Neumann heat kernel in the ball. J. Funct. Anal. **260**(2), 490–500 (2011)

[Pin80]   M.A. Pinsky, The eigenvalues of an equilateral triangle. SIAM J. Math. Anal. **11**(5), 819–827 (1980)

[Pin85]   M.A. Pinsky, Completeness of the eigenfunctions of the equilateral triangle. SIAM J. Math. Anal. **16**(4), 848–851 (1985)

[Pol]   Polymath7. The Hot Spots Conjecture, http://polymathprojects.org/2012/09/10/ polymath7-research-threads-4-the-hot-spots-conjecture/, Ch. Evans, T. Tao (moderators). Accessed 24 June 2012

[Pom75]   C. Pommerenke, *Univalent Functions* (Vandenhoeck & Ruprecht, Göttingen, 1975). With a chapter on quadratic differentials by Gerd Jensen, Studia Mathematica/Mathematische Lehrbücher, Band XXV

[Pom92]   Ch. Pommerenke. *Boundary Behaviour of Conformal Maps*. Grundlehren der Mathematischen Wissenschaften (Fundamental Principles of Mathematical Sciences), vol. 299 (Springer, Berlin, 1992)

[Pro90]   P. Protter, *Stochastic Integration and Differential Equations*. Applications of Mathematics (New York), vol. 21 (Springer, Berlin, 1990). A new approach

[PY86]   J. Pitman, M. Yor, Asymptotic laws of planar Brownian motion. Ann. Probab. **14**(3), 733–779 (1986)

[RY99]   D. Revuz, M. Yor, *Continuous Martingales and Brownian Motion*. Grundlehren der Mathematischen Wissenschaften (Fundamental Principles of Mathematical Sciences), 3rd edn., vol. 293 (Springer, Berlin, 1999)

[Shi85]   M. Shimura, Excursions in a cone for two-dimensional Brownian motion. J. Math. Kyoto Univ. **25**(3), 433–443 (1985)

[Sil74]   M.L. Silverstein, *Symmetric Markov Processes*. Lecture Notes in Mathematics, vol. 426 (Springer, Berlin, 1974)

[STW00]   F. Soucaliuc, B. Tóth, W. Werner, Reflection and coalescence between independent one-dimensional Brownian paths. Ann. Inst. H. Poincaré Probab. Stat. **36**(4), 509–545 (2000)

[Syt75]   G.N. Sytaja, Asymptotic representation of the probability of small deviation of the trajectory of a Brownian motion from a given function, in *Theory of Random Processes, No. 3 (Russian)*, pp. 117–121, 160 (Izdat. Naukova Dumka, Kiev, 1975)

[Syt77]   G.N. Sytaja, The asymptotic behavior of the Wiener measure of small spheres. Teor. Verojatnost. i Mat. Statist. **16**, 121–135, 157 (1977)

[Syt79]   G.N. Sytaja, On the problem of the asymptotic behavior of a Wiener measure of small spheres in the uniform metric, in *Analytical Methods of Probability Theory (Russian)*, pp. 95–98, 153 (Naukova Dumka, Kiev, 1979)

[VW85]   S.R.S. Varadhan, R.J. Williams, Brownian motion in a wedge with oblique reflection. Commun. Pure Appl. Math. **38**(4), 405–443 (1985)

[Wan94]   F.Y. Wang, Application of coupling methods to the Neumann eigenvalue problem. Probab. Theory Relat. Fields **98**(3), 299–306 (1994)

[Wik12a]   Wikipedia, Brownian motion—wikipedia, the free encyclopedia (2012), http://en.wikipedia.org/w/index.php?title=Brownian_motion&oldid=524637895. Accessed 24 Nov 2012

[Wik12b]   Wikipedia, Probabilistic method—wikipedia, the free encyclopedia (2012), http://en.wikipedia.org/w/index.php?title=Probabilistic_method&oldid=518260271. Accessed 24 Nov 2012

[Wik12c]   Wikipedia, Probabilistic proofs of non-probabilistic theorems—wikipedia, the free encyclopedia (2012), http://en.wikipedia.org/w/index.php?title=Probabilistic_proofs_of_non-probabilistic_theorems&oldid=475916464. Accessed 24 Nov 2012

LECTURE NOTES IN MATHEMATICS   🐴 Springer

Edited by J.-M. Morel, B. Teissier; P.K. Maini

Editorial Policy (for the publication of monographs)

1. Lecture Notes aim to report new developments in all areas of mathematics and their applications - quickly, informally and at a high level. Mathematical texts analysing new developments in modelling and numerical simulation are welcome.

   Monograph manuscripts should be reasonably self-contained and rounded off. Thus they may, and often will, present not only results of the author but also related work by other people. They may be based on specialised lecture courses. Furthermore, the manuscripts should provide sufficient motivation, examples and applications. This clearly distinguishes Lecture Notes from journal articles or technical reports which normally are very concise. Articles intended for a journal but too long to be accepted by most journals, usually do not have this "lecture notes" character. For similar reasons it is unusual for doctoral theses to be accepted for the Lecture Notes series, though habilitation theses may be appropriate.

2. Manuscripts should be submitted either online at www.editorialmanager.com/lnm to Springer's mathematics editorial in Heidelberg, or to one of the series editors. In general, manuscripts will be sent out to 2 external referees for evaluation. If a decision cannot yet be reached on the basis of the first 2 reports, further referees may be contacted: The author will be informed of this. A final decision to publish can be made only on the basis of the complete manuscript, however a refereeing process leading to a preliminary decision can be based on a pre-final or incomplete manuscript. The strict minimum amount of material that will be considered should include a detailed outline describing the planned contents of each chapter, a bibliography and several sample chapters.

   Authors should be aware that incomplete or insufficiently close to final manuscripts almost always result in longer refereeing times and nevertheless unclear referees' recommendations, making further refereeing of a final draft necessary.

   Authors should also be aware that parallel submission of their manuscript to another publisher while under consideration for LNM will in general lead to immediate rejection.

3. Manuscripts should in general be submitted in English. Final manuscripts should contain at least 100 pages of mathematical text and should always include

   - a table of contents;
   - an informative introduction, with adequate motivation and perhaps some historical remarks: it should be accessible to a reader not intimately familiar with the topic treated;
   - a subject index: as a rule this is genuinely helpful for the reader.

   For evaluation purposes, manuscripts may be submitted in print or electronic form (print form is still preferred by most referees), in the latter case preferably as pdf- or zipped ps-files. Lecture Notes volumes are, as a rule, printed digitally from the authors' files. To ensure best results, authors are asked to use the LaTeX2e style files available from Springer's web-server at:

   ftp://ftp.springer.de/pub/tex/latex/svmonot1/ (for monographs) and
   ftp://ftp.springer.de/pub/tex/latex/svmultt1/ (for summer schools/tutorials).

Additional technical instructions, if necessary, are available on request from lnm@springer.com.

4. Careful preparation of the manuscripts will help keep production time short besides ensuring satisfactory appearance of the finished book in print and online. After acceptance of the manuscript authors will be asked to prepare the final LaTeX source files and also the corresponding dvi-, pdf- or zipped ps-file. The LaTeX source files are essential for producing the full-text online version of the book (see http://www.springerlink.com/openurl.asp?genre=journal&issn=0075-8434 for the existing online volumes of LNM). The actual production of a Lecture Notes volume takes approximately 12 weeks.

5. Authors receive a total of 50 free copies of their volume, but no royalties. They are entitled to a discount of 33.3 % on the price of Springer books purchased for their personal use, if ordering directly from Springer.

6. Commitment to publish is made by letter of intent rather than by signing a formal contract. Springer-Verlag secures the copyright for each volume. Authors are free to reuse material contained in their LNM volumes in later publications: a brief written (or e-mail) request for formal permission is sufficient.

Addresses:
Professor J.-M. Morel, CMLA,
École Normale Supérieure de Cachan,
61 Avenue du Président Wilson, 94235 Cachan Cedex, France
E-mail: morel@cmla.ens-cachan.fr

Professor B. Teissier, Institut Mathématique de Jussieu,
UMR 7586 du CNRS, Équipe "Géométrie et Dynamique",
175 rue du Chevaleret
75013 Paris, France
E-mail: teissier@math.jussieu.fr

For the "Mathematical Biosciences Subseries" of LNM:

Professor P. K. Maini, Center for Mathematical Biology,
Mathematical Institute, 24-29 St Giles,
Oxford OX1 3LP, UK
E-mail: maini@maths.ox.ac.uk

Springer, Mathematics Editorial, Tiergartenstr. 17,
69121 Heidelberg, Germany,
Tel.: +49 (6221) 4876-8259

Fax: +49 (6221) 4876-8259
E-mail: lnm@springer.com